Synthesis Lectures on Communications

This series of short books cover a wide array of topics, current issues, and advances in key areas of wireless, optical, and wired communications. The series also focuses on fundamentals and tutorial surveys to enhance an understanding of communication theory and applications for engineers.

Isiaka Alimi

5G Fixed Wireless Access

Revolutionizing Connectivity
in the Digital Age

Isiaka Alimi
Instituto de Telecomunicações
University of Aveiro
Aveiro, Portugal

ISSN 1932-1244　　　　　　ISSN 1932-1708　(electronic)
Synthesis Lectures on Communications
ISBN 978-3-031-77538-3　　　ISBN 978-3-031-77539-0　(eBook)
https://doi.org/10.1007/978-3-031-77539-0

© The Editor(s) (if applicable) and The Author(s), under exclusive license to Springer
Nature Switzerland AG 2025

This work is subject to copyright. All rights are solely and exclusively licensed by the Publisher, whether the whole or part of the material is concerned, specifically the rights of translation, reprinting, reuse of illustrations, recitation, broadcasting, reproduction on microfilms or in any other physical way, and transmission or information storage and retrieval, electronic adaptation, computer software, or by similar or dissimilar methodology now known or hereafter developed.
The use of general descriptive names, registered names, trademarks, service marks, etc. in this publication does not imply, even in the absence of a specific statement, that such names are exempt from the relevant protective laws and regulations and therefore free for general use.
The publisher, the authors and the editors are safe to assume that the advice and information in this book are believed to be true and accurate at the date of publication. Neither the publisher nor the authors or the editors give a warranty, expressed or implied, with respect to the material contained herein or for any errors or omissions that may have been made. The publisher remains neutral with regard to jurisdictional claims in published maps and institutional affiliations.

This Springer imprint is published by the registered company Springer Nature Switzerland AG
The registered company address is: Gewerbestrasse 11, 6330 Cham, Switzerland

If disposing of this product, please recycle the paper.

To my loving wife, Abiola, whose unwavering support and belief in me made this journey possible.

To my wonderful children, Fatima, Ibrahim, Maryam, and Ummuani, who inspire me every day with their boundless curiosity, love, and joy.

And to the memory of my parents, who instilled in me the love of learning and the courage to chase my dreams. Your guidance and love continue to inspire me every day.

... This book is for all of you, with all my heart.

Preface

The COVID-19 pandemic has been a pivotal moment for technological progress and has underscored the critical importance of home-based services. It particularly highlighted the vital need for dependable, high-speed broadband connections to facilitate remote work, maintain social connections, and enable online learning, demonstrating that broadband access (BBA) is essential for socio-economic development. As a result, the demand for robust, dependable, adaptable, and secure home broadband increased exponentially. Furthermore, the worldwide need for residential broadband is expected to stay strong after the pandemic, with projections indicating it will surpass 58 million subscribers by 2026. In 2020, during the pandemic, the global residential broadband market expanded to more than 1.1 billion subscribers, marking a 4% increase from the year before.

The increasing growth rate is also due to the substantial rise in mobile data traffic demand and higher access rates over the past decade, driven by innovative services and applications such as financial trading, autonomous vehicle systems, and Internet-of-Things (IoT) devices. These services necessitate not just higher bandwidth, but also ultra-low latency, exceptional reliability, minimal energy consumption, and massive connection capacity. Consequently, telecommunication operators should reassess their access infrastructure and improve their management practices to achieve better network performance and user experience, all while maintaining cost and energy efficiency.

The Fifth Generation (5G) cellular technology stands as the most significant advancement of this decade. Introduced as a revolutionary broadband technology, 5G addresses network requirements and enhances the quality of service for both mobile and residential users. Its introduction is accelerating global radio access network (RAN) densification, fostering new network designs, and enabling innovative applications with rigorous performance standards, such as reduced latency, higher throughput, and increased reliability. Consequently, 5G offers enhanced capacity and connectivity, supporting advanced applications and services that demand ultra-high connection speeds, traffic handling, and mobility. This technology will influence various aspects of daily life, including knowledge sharing, connectivity, entertainment, and safety. It supports a wide range of use cases, including smart city integration, autonomous vehicles, augmented reality, medical advancements, new forms of social interaction, and improved fixed wireless connectivity.

Optical access networks have been providing both residential and mobile services for the final kilometers via various fiber topologies. However, deploying wireline broadband networks entails significant capital investment and a commitment to mid to long-term funding. The deployment of the last few hundred meters, known as the last mile, has been particularly challenging for wireline broadband technologies. A time-efficient and cost-effective alternative for last-mile deployment is Fixed Wireless Access (FWA). This innovative approach facilitates the quick and easy rollout of broadband networks. With the advent of 5G, FWA is set to become the fastest-growing residential broadband access technology. 5G FWA enables carriers to address the growing demand for high-speed broadband services more efficiently and with faster time-to-market solutions.

This book, *Fixed Wireless Access: Revolutionizing Connectivity in the Digital Age*, aims to comprehensively explore FWA, tracing its historical development, examining its current applications, and assessing its future potential. The book begins with a thorough examination of the foundational principles of FWA, providing readers with an in-depth understanding of its components, technologies, and deployment strategies. It then delves into the technical intricacies, market dynamics, and real world applications that underscore the versatility and significance of FWA. In this context, it explores the transformative potential of 5G technology in delivering highspeed broadband services. So, several key areas, including high-frequency spectrum bands, advanced transmission schemes, multi-connectivity, adaptive numerology, and Integrated Access and Backhaul (IAB), among other things, which are critical for enhancing network performance, increasing capacity, and reducing latency, are covered. Beyond offering technical insights, the book highlights the socio-economic impact of FWA. By enabling cost-effective connectivity in remote and underserved areas, FWA has the potential to narrow or bridge the digital divide, fostering inclusivity and providing access to education, healthcare, and economic opportunities for all. This makes the book a crucial resource for grasping the future of broadband connectivity and the strategic approaches required to tackle deployment challenges. This book is intended to be a definitive guide for professionals, researchers, policymakers, and anyone interested in understanding the nuances and implications of this transformative technology.

Aveiro, Portugal
July 2024

Isiaka Alimi
iaalimi@ua.pt

Acknowledgements

The preparation of this book was supported by the Fundação para a Ciência e a Tecnologia (FCT). The author gratefully acknowledges the support of the FCT under the grant agreement "Ultra-High-Capacity Optical Wireless Communication with Low-Complexity Transceiver for Effective Spectrum Utilization in Beyond-5G Networks," DOI 10.54499/2021.03815.CEECIND/CP1653/CT0005 (https://doi.org/10.54499/2021.03815.CEECIND/CP1653/CT0005). Acknowledgment is also extended to the European Regional Development Fund (FEDER) through the Competitiveness and Internationalization Operational Programme (CENTRO 2020) for the ORCIP project, under Grant CENTRO-01-0145-FEDER-022141. Additionally, the technical support provided by the Marie Skłodowska-Curie Actions (MSCA) through the Horizon 2020 Research and Innovation Staff Exchange (RISE) programme, specifically the Deep Intelligent Optical and Radio Communication Networks (DIOR) project under Grant Agreement ID: 101008280, is deeply appreciated.

Contents

1	**Introduction**		1
	1.1 Overview		1
	1.2 Broadband System		1
		1.2.1 Use Cases Classification	2
		1.2.2 Home Broadband	2
		1.2.3 Primary Issues and Potential Solutions	2
	1.3 Fixed Wireless Broadband Overview		5
		1.3.1 Fixed Broadband Technologies	6
		1.3.2 Fixed Wireless Access Network	14
	1.4 5G FWA Opportunities		16
		1.4.1 Improved Broadband Speed	17
		1.4.2 Fast Network Development	17
		1.4.3 Homogenized Wider Coverage	18
		1.4.4 Flexibility	18
		1.4.5 Close the Digital Divide	18
	1.5 FWA Technical Challenges and Solutions		19
		1.5.1 Spectrum-Based	19
		1.5.2 Deployment-Based	19
		1.5.3 Interoperability	20
		1.5.4 Network Congestion	21
	1.6 FWA Implementation Considerations		21
		1.6.1 Spectrum	21
		1.6.2 Sites	24
		1.6.3 Capacity	24
		1.6.4 Service Speeds	25
		1.6.5 Devices	27
	1.7 Conclusion		28
	References		28

2 Network Architecture and Evolution ... 35
- 2.1 Introduction ... 35
- 2.2 Wireless Next-Generation Technologies ... 36
  - 2.2.1 5G New Radio ... 36
  - 2.2.2 5G Use Cases ... 37
- 2.3 RAN Architecture ... 38
  - 2.3.1 3GPP RAN Architecture ... 39
  - 2.3.2 ITU-T RAN Architecture ... 44
  - 2.3.3 Virtualized RAN ... 45
  - 2.3.4 Open RAN ... 45
- 2.4 Conclusion ... 46
- References ... 46

3 Transport Network Architectures and Requirements ... 51
- 3.1 Introduction ... 51
- 3.2 Transport Network Architecture ... 51
  - 3.2.1 Core Network ... 52
  - 3.2.2 Metro Network ... 52
  - 3.2.3 Access Network ... 52
- 3.3 Transport Network Architecture Evolution ... 53
- 3.4 FWA Transport Network ... 53
  - 3.4.1 Fronthaul Network Interfaces ... 54
  - 3.4.2 RAN Split Options for FWA ... 56
- 3.5 Conclusion ... 57
- References ... 58

4 5G FWA Technological Improvements ... 61
- 4.1 Introduction ... 61
- 4.2 Radio Access Network ... 62
  - 4.2.1 Increased Spectral Efficiency ... 62
  - 4.2.2 New Spectrum ... 62
  - 4.2.3 Beam Management Frameworks for 5G NR ... 63
  - 4.2.4 Improved Channel State Information Mechanisms ... 64
  - 4.2.5 Multi-antenna Systems ... 65
  - 4.2.6 Downlink Coverage Utilization Expansion ... 66
  - 4.2.7 Uplink Performance Improvement ... 66
  - 4.2.8 Service Differentiation ... 66
  - 4.2.9 Higher Order Modulation ... 67
  - 4.2.10 Dynamic TDD and Interference Mitigation ... 67
- 4.3 5G Core Network ... 67
- 4.4 Conclusion ... 68
- References ... 68

5	**5G FWA Customer Premise Equipment**		73
	5.1 Introduction		73
	5.2 FWA and Mobile CPEs		73
		5.2.1 Physical Size	74
		5.2.2 TX Power/EIRP Limitations	74
		5.2.3 Antennas and Advanced Features	75
		5.2.4 Battery Life Issues	75
	5.3 CPE Solutions		75
		5.3.1 Indoor CPE	75
		5.3.2 Outdoor CPE	76
		5.3.3 Flexi CPE	76
	5.4 5G NR CPE Power Classes		77
		5.4.1 FR1 UE Power Classes	77
		5.4.2 FR2 UE Power Classes	78
	5.5 FWA CPE Performance, Capabilities, and Applications		78
	5.6 Conclusion		81
	References		81
6	**Wireless Network Coverage Planning**		85
	6.1 Introduction		85
		6.1.1 Fundamentals of Radio Propagation	85
		6.1.2 Link Planning Tools	87
	6.2 AI-Based Link Planning Tools		88
		6.2.1 Evolutionary Algorithms for Network Planning	88
		6.2.2 Machine Learning-Based 3D RF Planning Tool	89
		6.2.3 AI-Based Computer Vision Aided Network Coverage Planning	89
	6.3 Conclusion		90
	References		91
7	**Key Methods for Efficient and High-Speed FWA Solutions**		95
	7.1 Introduction		95
	7.2 Millimeter-Wave and Enabling FWA Solutions		96
		7.2.1 MmWave-Only SA	96
		7.2.2 Extended-Range MmWave 5G FWA	96
	7.3 Advancement of Network Nodes		100
		7.3.1 Integrated Access and Backhaul	100

	7.3.2	Repeater	106
	7.3.3	Intelligent Reflective Surface	107
7.4	Transmission Schemes Convergence		108
7.5	Multi-X Connectivity		108
	7.5.1	5G Carrier Aggregation and Dual-Connectivity	109
	7.5.2	Decoupled Uplink and Downlink Access	111
7.6	Adaptive 5G Numerology and Multi-layer Spectrum Management		112
	7.6.1	Multi-layer 5G Spectrum	112
	7.6.2	5G Flexible Numerology	113
7.7	Conclusion		115
References			115

Conclusion 119

Acronyms

3GPP	Third-Generation Partnership Project
4G	Fourth-Generation
5G	Fifth-Generation
AI	Artificial Intelligence
APs	Access Points
AR	Augmented Reality
AUs	Access Units
B5G	Beyond 5G
BBU	Baseband Unit
bps	Bits Per Second
BSs	Base Stations
CA	Carrier Aggregation
CAPEX	Capital Expenditure
CC	Component Carrier
CDMA	Code-Division Multiple Access
CM	Cable Modem
CMTS	Cable Modem Termination System
CNNs	Convolutional Neural Networks
CO	Central Office
CoMP	Coordinated Multi-Point
COTS	Commercial off-the-Shelf
COVID-19	Coronavirus Disease 2019
CPE	Customer Premises Equipment
CPRI	Common Public Radio Interface
CQI	Channel Quality Indicator
CRA	Contention Resolution Algorithm
C-RAN	Cloud Radio Access Network
CSI-RS	Channel State Information Reference Signal

CSPs	Communication Service Providers
CU	Central Unit
DAG	Directed Acyclic Graph
dB	Decibels
DC	Dual Connectivity
DeNB	Donor evolved Node B
DL	Downlink
DNNs	Deep Neural Networks
DOCSIS	Data Over Cable Service Interface Specification
DPUs	Distributed Point Units
D-RAN	Distributed RAN
DRX	Discontinuous Reception
DS	Downstream
DSL	Digital Subscriber Line
DSM	Dynamic Spectrum Management
DU	Distributed Unit
E2E	End-to-End
eCPRI	Enhanced CPRI
EIRP	Effective Isotropic Radiated Power
eMBB	Enhanced MBB
eNB	Evolved Node B
E-UTRA	Evolved-Universal Terrestrial Radio Access
FCC	Federal Communications Commission
FCN	Fully Convolutional Network
FDD	Frequency-Division Duplex
FEC	Forward Error Correction
FER-CNN	Faster Edge Region CNN
FR1	Frequency Range 1
FR2	Frequency Range 2
FSAN	Full Service Access Network
FSO	Free-Space Optics
FTTB	Fibre-to-the-Building
FTTC	Fibre-to-the-Cabinet
FTTH	Fibre-to-the-Home
FTTN	Fiber-to-the-Node
FTTP	Fibre-to-the-Premise
FTTx	Fibre-to-the-x
FWA	Fixed Wireless Access
Gbps	Gigabits per second
GDP	Gross Domestic Product
GHP	Guaranteed Household Percentage

GIS	Geographical Information System
gNB	Next generation Node B
GPON	Gigabit Passive Optical Network
GPUs	Graphical Processing Units
HBB	Home Broadband
HetNets	Heterogeneous Networks
HFC	Hybrid Fiber-Coaxial
HLS	High-Level Split
HSDPA	High-Speed Downlink Packet Access
IAB	Integrated Access and Backhaul
IEEE	Institute of Electrical and Electronics Engineers
IMS	IP Multimedia Subsystem
IoT	Internet of Things
IPTV	Internet Protocol Television
IRS	Intelligent Reflective Surfaces
ISD	Inter-Site Distance
ISDN	Integrated Services Digital Network
ISPs	Internet Service Providers
ITU-T	International Telecommunication Union Telecommunication Standardization Sector
JP	Joint Processing
L1	Layer 1
L2	Layer 2
L3	Layer 3
LDCs	Least Developed Countries
LEO	Low Earth Orbit
LLS	Low-Level Split
LLX	Low Latency xHaul
LoS	Line-of-Sight
LPWA	Low Power Wide Area
LTE	Long-Term Evolution
LTE-A	LTE-Advanced
MAC	Media Access Control
MBB	Mobile Broadband
MBS	Main BS
MC	Multi-Connectivity
MCS	Modulation and Coding Schemes
MDUs	Multi-Dwelling Units
MEC	Multi-Access Edge Computing
MFH	Mobile Fronthaul
MIMO	Multiple-Input Multiple-Output

mMTC	Massive Machine Type Communications
mmWave	Millimeter Wave
MNOs	Mobile Network Operators
MPE	Maximum Permissible Exposure
MR-DC	Multi-Radio Dual Connectivity
MRRUs	Macro-cell RRUs
mRRUs	Micro-cell RRUs
MU-MIMO	Multi-User MIMO
MVNOs	Mobile Virtual Network Operators
NFV	Network Function Virtualization
NG-EPON	Next-generation EPON
NGFI	Next Generation Fronthaul Interface
NG-RAN	Next Generation RAN
NN	Neural Networks
NR	New Radio
NR-DC	NR Dual Connectivity
NSA	Non-Standalone
ODN	Optical Distribution Network
OECD	Organisation for Economic Co-operation and Development
OFDM	Orthogonal Frequency-Division Multiplexing
OLT	Optical Line Terminal
ONTs	Optical Network Terminals
ONUs	Optical Network Units
OPEX	Operating Expenses
OWC	Optical Wireless Communications
P2MP	Point-to-Multipoint
P2P	Point-to-Point
PCell	Primary Cell
PDCP	Packet Data Convergence Protocol
PHY	Physical
PL	Path Loss
PON	Passive Optical Network
POTS	Plain Old Telephone Service
PRINS	Protocol for N32 Interconnect Security
PS	Protocol Stack
PSO	Particle Swarm Optimization
PSs	Phase Shifters
QoE	Quality of Experience
QoS	Quality of Service
RAN	Radio Access Network
RAT	Radio Access Technology

R-CNN	Region-based CNN
RE	Radio Equipment
RF	Radio Frequency
RFFE	RF Front-Ends
RLC	Radio Link Control
RN	Relay Node
RoE	Radio over Ethernet
ROI	Return On Investment
RoW	Right-of-Way
RRC	Radio Resource Control
RRH	Remote Radio Head
RRU	Remote Radio Unit
RSSI	Received Signal Strength Indicator
RU	Radio Unit
SA	Standalone
SBA	Service-Based Architecture
SBSs	Secondary BSs
SCC	Secondary Component Carriers
SCell	Secondary Cell
SDN	Software-Defined Networking
SINR	Signal-to-Interference-plus-Noise Ratio
SLA	Service-Level Agreement
SMEs	Small and Medium-sized Enterprises
SMF	Single-Mode Fiber
SRS	Sound Reference Signal
SRS	Sound Reference Signal
ST	Spanning Tree
SU-MIMO	Single-User MIMO
Tbps	Terabits per second
TC	Transmission Convergence
TCO	Total Cost of Ownership
TDD	Time-Division Duplex
TDM	Time-Division Multiplexing
TLS	Transport Layer Security
TRP	Total Radiated Power
TSN	Time-Sensitive Networking
UE	User Equipment
UHD	Ultra High Definition
UL	Uplink
uRLLC	Ultra-Reliable Low Latency Communications
US	Upstream

UTP	Unshielded Twisted Pairs
VoIP	Voice over Internet Protocol
VR	Virtual Reality
vRAN	Virtualized RAN
WDM	Wavelength-Division Multiplexing
Wi-Fi	Wireless Fidelity
WiMAX	Worldwide Interoperability for Microwave Access
WISPs	Wireless Internet Service Providers
WWW	World Wide Web

Introduction 1

1.1 Overview

Home broadband (HBB) has become crucial for contemporary living, enabling remote work, streaming entertainment, online education, and smart home management, leading to a substantial global demand for enhanced broadband connectivity. In addition to over one billion homes still lacking connection and many others being underserved [1], challenges regarding the availability and speed of broadband services persist, contributing to a significant *digital divide*. This digital divide highlights the disparity between affluent and impoverished communities, as well as between regions and populations with access to modern information and communication technology and those with limited or no access. The *Broadband for All* initiative seeks to bridge this gap by providing high-speed broadband access to everyone, everywhere, thereby connecting all individuals to the digital economy [2].

1.2 Broadband System

Broadband refers to the transmission of a large amount of data over high-speed Internet connections that are constantly active. Delivered through various technologies that depend on location, broadband became prevalent in the late 1990s, largely replacing older dial-up connections that utilized the audible frequencies of telephone lines [3–6].

The coronavirus disease 2019 (COVID-19) pandemic highlighted the crucial importance of reliable high-speed broadband connections to support remote work and online learning. This demand has persisted beyond the pandemic, with communication and conferencing platforms like Cisco WebEx, Zoom, Apple FaceTime Zoom, and Microsoft Teams; and live streaming applications such as Google meet, PPLive, Facebook, YouTube, and Instagram; seeing a substantial rise in usage. To accommodate this ongoing need, HBB must be robust, flexible, reliable, and secure. Various transport technologies can be used for HBB, including cable, Digital Subscriber Line (DSL), fiber optics, Ethernet, 4G/5G, satellite, or fixed wireless connections [7–9]. This section, based on use case classification, outlines the current state of broadband systems, highlights key issues, and explores potential solutions.

1.2.1 Use Cases Classification

The increasing demand for high-speed, affordable Internet for residential users has spurred the development of Fixed Wireless Access (FWA) technologies leveraging 5G New Radio (NR). Broadband-based services can be provided to homes through indoor or outdoor subscriber units connected to the wireless network via 5G NR technology. Coverage within a home can be enhanced with additional devices. Residential use cases are typically divided into two categories: Rural and Suburban/Urban [7].

Geographies are categorized as urban or rural according to population size and density. Urban areas are characterized by dense development and include commercial, residential, and other non-residential uses. In contrast, rural areas encompass all land, population, and housing not classified as urban. According to this classification, approximately 80% of the population is considered urban, while the other 20% is classified as rural [10, 11].

In rural regions, the population is spread across a significantly larger geographic area, leading to significantly lower population densities compared to urban areas. Conversely, urban areas have a much higher median population density. These differences in population distribution impact the types of fixed broadband technologies that Internet Service Providers (ISPs) deploy to serve rural and urban consumers [12, 13].

1.2.2 Home Broadband

Digital divides are evident in both the availability and utilization of broadband communications. Developed countries have more extensive broadband and Internet subscription and penetration rates compared to Least Developed Countries (LDCs). Individuals in Europe, affluent North American countries, and certain parts of Asia and the Pacific are more likely to have Internet access compared to people in other regions, especially Africa. Evidence indicates that the disparity between LDCs and other developing nations is widening, which raises apprehensions about the potential negative impact on efforts to achieve the Sustainable Development Goals [2, 6, 12, 14].

1.2.3 Primary Issues and Potential Solutions

The digital divide extends beyond just the insufficient Internet access. Even households with broadband access still encounter difficulties associated with the emerging digital divide.

As technological advancements and social development progress, an increasing amount of content is delivered via broadband, making the speed and capabilities of broadband services increasingly crucial. For instance, while about 96% of households in the UK have access to high-speed broadband, only 18% are connected with full fiber [15]. In 2022, it was estimated that between 23 and 25 million Americans are either unserved or underserved by broadband Internet, highlighting the ongoing challenge of ensuring widespread access to the Internet and its services [16, 17]. The World Bank reports that a 10% rise in broadband penetration can lead to a 1.3% rise in average Gross Domestic Product (GDP) and up to a 3% increase in job creation. However, 6 billion people still do not have access to high-speed Internet, and 3.8 billion lack any Internet access whatsoever [2]. The following subsections address some of the primary issues contributing to the digital divide and broadband challenges, as well as present a multifaceted approach to ensuring reliable and high-speed broadband access for everyone.

1.2.3.1 Geographic Disparities

Typically, urban areas benefit from superior access to high-speed broadband due to denser infrastructure and greater competition among providers. In contrast, rural and remote regions encounter considerable difficulties in securing reliable broadband. The high cost of deploying infrastructure in these sparsely populated areas leads to slower speeds and less reliable service. Even within suburban areas, broadband speeds can vary significantly depending on infrastructure investment and provider competition [7, 12, 15]. Collaborative efforts among governments, private companies, and community organizations are essential for developing more effective broadband solutions.

1.2.3.2 Infrastructure Limitations

Many regions continue to depend on outdated infrastructure, like copper lines, which are inadequate for supporting modern high-speed Internet needs. Additionally, the rollout of fiber optic cables—known for their superior speed and reliability—is inconsistent. While some areas benefit from advanced fiber connections, others remain reliant on older, slower technologies [2, 15, 16]. To tackle these challenges, it is essential to increase investment in modern broadband infrastructure, especially in rural and underserved areas.

1.2.3.3 Economic Barriers

Disparities in service costs and investment are significant barriers to achieving Broadband for All. For instance, the high costs associated with high-speed broadband—such as monthly fees, installation charges, and equipment rentals—can make it inaccessible for many households, especially those with lower incomes. Additionally, broadband providers may focus on more profitable urban and suburban areas, neglecting rural regions and thereby widening the digital divide [2]. Implementing initiatives to reduce broadband costs for low-income households can help ensure more equitable access [6].

1.2.3.4 Technological Challenges

Bandwidth constraints can lead to slow speeds during peak usage periods, while latency issues can affect applications that depend on real-time data transmission, including online gaming and video conferencing. Although FWA and satellite broadband provide alternative solutions, they frequently encounter challenges such as signal interference, weather sensitivity, and higher latency compared to wired connections. To address these issues, ongoing innovation in broadband technologies, including improvements in FWA, satellite, and fiber optics, will be crucial [18–20].

1.2.3.5 Performance Variability

The broadband speeds advertised often differ from the actual performance experienced by users. Many people may find that their speeds are slower than promised, especially during peak usage times. Additionally, fluctuations in Quality of Service (QoS), such as outages and slowdowns, can affect user satisfaction and productivity [21–23].

1.2.3.6 Emerging Demands

The increasing demand for data-heavy applications like 4K streaming, cloud computing, and Internet of Things (IoT) devices adds extra pressure on existing broadband networks. The COVID-19 pandemic has intensified the demand for dependable high-speed Internet as more people work and study from home. This shift has underscored the shortcomings of current broadband infrastructure [8, 24–27].

1.2.3.7 Regulatory and Policy Issues

Regulatory frameworks play a crucial role in shaping broadband deployment and competition. Policies that do not encourage investment in underserved areas can impede progress. Additionally, the ongoing debate over net neutrality influences how ISPs handle traffic and prioritize different types of data, which can affect overall Internet speed and accessibility [24]. Therefore, supportive policies and regulations that foster competition and incentivize expansion into less profitable regions are essential for bridging the digital divide [6, 14, 28, 29].

1.3 Fixed Wireless Broadband Overview

As the World Wide Web (WWW) extended beyond academic and military use, the demand for home data connections started to emerge. Early Internet access was facilitated by modems operating at speeds of 14.4, 28.8, and 56.6 kbps over the Plain Old Telephone Service (POTS), offering dial-up connections through various service providers. Also, for business customers, Integrated Services Digital Network (ISDN) emerged as an *always-on* alternative to conventional dial-up connections, offering service speeds of 64 and 128 kbps. As demand grew with the increasing use of emails, attachments, and network servers, Frame Relay and dedicated T1 lines, despite their higher costs, became more common solutions. Also, as the demand for always-on broadband increased in the residential sector, the need for higher throughput prompted the development of technologies such as the DSL family (xDSL). These innovations were designed to boost download speeds beyond 1 Mbps [6, 30].

Furthermore, different wireless access solutions were explored, such as satellite access, point-to-point (P2P) Wireless Fidelity (Wi-Fi) for extended coverage, and emerging technologies like Worldwide Interoperability for Microwave Access (WiMAX), which is based on the Institute of Electrical and Electronics Engineers (IEEE) 802.16 standards. Primarily designed for the 10–66 GHz frequency range, WiMAX technology was later adapted to lower frequencies starting at 2 GHz. The introduction of Multiple-Input Multiple-Output (MIMO) and Orthogonal Frequency-Division Multiplexing (OFDM) technologies further enhanced its capabilities. As Long-Term Evolution (LTE) standards were established, the performance of LTE surpassed that of WiMAX and alternative technologies. In this regard, LTE-Time-Division Duplex (TDD) emerged as a viable solution for FWA, enhancing its reach and effectiveness. As discussed in Chap. 2, the advancements in transmission techniques with 5G NR, such as increased channel bandwidth and subcarrier spacing, will enable the expansion of FWA deployments in rural areas. These improvements will enable FWA to offer greater bandwidth more effectively than fiber [7, 31].

Global need for broadband connections has been robust for numerous years. The broadband market is divided into mobile and fixed segments. In terms of global subscriptions, the mobile broadband (MBB) market significantly surpasses the fixed broadband counterpart, with around 6 billion MBB subscriptions compared to 1.18 billion fixed broadband subscriptions as of November 2020. Despite this, the fixed broadband market continues to grow worldwide [7]. For instance, recent Organisation for Economic Co-operation and Development (OECD) statistics reveal that Fiber and FWA have experienced the most significant growth among fixed broadband technologies over the past three years. In this context, fiber subscriptions have risen by 56% from June 2020 to June 2023, while FWA subscriptions have grown by 64% [32].

The increasing number of devices and their rising usage are leading wireless subscribers to demand more bandwidth and ubiquitous connectivity, placing greater pressure on wireless networks. Therefore, it is essential to improve network performance, particularly in terms of capacity and coverage, especially in dynamic environments such as densely populated areas. In the initial phase of the 5G network rollout, Mobile Network Operators (MNOs) have primarily utilized the mid-band spectrum combined with massive MIMO to boost network capacity. Small cell network densification has been suggested as the next logical advancement for a fast, smart, and cost-effective approach to improving reliability, boosting capacity, expanding coverage, and offloading traffic from macro networks amid a data surge. By densifying the network to increase capacity, mobile bandwidth can be improved, providing end users with a seamless mobile experience [33–35]. The following section offers a comprehensive overview of both emerging and established wireline transport technologies applicable for 5G network densification.

1.3.1 Fixed Broadband Technologies

With the advent of the 5G era, expectations are high. For instance, Industry 4.0 and the evolving digital society will require networks that offer low latency and high throughput across all areas. Consequently, both fixed and mobile networks will need to be expanded and enhanced to effectively support a range of consumer and business applications [6, 25].

While existing dark fiber assets have effectively facilitated LTE macro site deployments, the densification needs for 5G mm-Wave and mid-band will necessitate unmatched amounts of additional fiber infrastructure for 5G xHaul, encompassing fronthaul, midhaul, and backhaul [9]. Fiber densification has been expensive to implement, particularly when it comes to connecting customer premises (Fiber to the Home), often referred to as the *last mile*, owing to the extensive civil work required and the need to coordinate with various stakeholders (such as property, duct, road, and pavement owners). Also, subjective evidence suggests that the final hundred meters can account for about 90% of the costs. Similar financial pressures are also observed when expanding and upgrading wireline infrastructure from legacy copper to advanced fiber networks [25, 36].

The primary technologies for delivering fixed broadband are xDSL, cable, and fiber, each emerging as viable and strategically important alternatives to dark fiber transport for 5G [7, 9]. The most common fixed broadband connections are illustrated in Fig. 1.1. Also, Table 1.1 outlines the benefits and drawbacks of these three technologies. Additionally, recent advancements in Ethernet are expected to streamline and lower the expenses and complexities associated with wireless transport deployments, proving crucial for wireless densification [9]. Similarly, Free-Space Optics (FSO) systems provide low-cost, ultra highspeed wireless connectivity for various last-mile applications. This technology offers service providers with an affordable, high-bandwidth wireless solution to extend backbone networks across multiple buildings, eliminating the need for fiber-optic cable infrastructure or Right-of-Way (RoW) [27, 37, 38].

In rural areas, satellite-based connectivity is the predominant method for providing Internet access, followed by FWA. FWA can offer an economical approach for network densification. It also offers a more accessible and affordable way to provide broadband connectivity

1.3 Fixed Wireless Broadband Overview

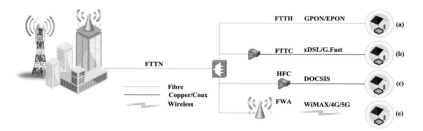

Fig. 1.1 Fixed broadband technologies

Table 1.1 Main fixed broadband technologies

	xDSL	Cable	Fiber
Pros	Relatively low investment if copper infrastructure is already in place	Relatively low investment in passive equipment using existing CATV infrastructure	Extremely high speeds
Cons	• Speeds are limited and depend on the length of the copper • Highly asymmetrical • Significant investment in active equipment • Rising OPEX • Restricted availability of copper	• Marginally greater speeds than xDSL, though reliant on distance • Shared bandwidth • Being unable to unbundle services restricts competition • Limited global availability, especially in areas with a digital divide	• Extremely high upfront investment costs, including for civil engineering and ducting • Development challenges with permitting frequently lead to delays in time to market
Future-proof	Low	Medium	High
Typical speeds	• 20–100 Mbps DL • 1–40 Mbps UL	• 200+ Mbps DL • 100+ Mbps UL	• 1 Gbps DL (and more) • 1 Gbps UL (and more)

DSL Digital Subscriber Line; *OPEX* Operating Expense; *CATV* Cable Television; *UL*Uplink; *DL* Downlink

in areas lacking wireline infrastructure or where only copper wireline infrastructure is available. Also, when factors like household density and civil work costs make fiber deployment advantageous, FWA can be utilized to complement fixed broadband installations [25]. Also, urban environments typically have greater cable and fiber coverage, resulting in faster speeds compared to rural areas. It is imperative to highlight that fixed wireless technology serves

Table 1.2 Main cost drivers

Technologies	Primary cost components	Drivers
FTTH	Civil works	Characteristics of the local terrain, local regulations
	Fibre cable	Total length needed to reach all locations/premises
	PON splitters and split ratio (1:x)	Number of locations covered
	Local exchange equipment	Number of covered locations
	Last drop to customer premises (e.g. ONT & CPE)	Number of subscribers and the availability of infrastructure at customer sites
DSL	G.Fast DPUs	Coverage (range of copper sub-loop) and capacity (number of ports)
	Fibre cable and infrastructure	Local terrain features, regional regulations, distance to the DPUs to be served, and the number of premises to be served
	CPE	Number of subscribers
HFC	Civil works	Local terrain features and regional regulations
	Coaxial cable	Total length to reach all premises
	Active equipment (taps, splitters, amplifiers, and optical nodes)	Based on the length of the coaxial cable and the number of premises covered
	Fibre cable and infrastructure	Local terrain features, regional regulations, distance to optical nodes, and the number of premises to be covered
	Last drop to customer premises (e.g. CPE)	Subscriber count and availability of infrastructure at customer locations
FSO	Passive infrastructure (Shelter, streetlamp/utility poles)	Fixed costs based on coverage and capacity requirements
	Active infrastructure (e.g., Optical source, optical transceiver units and software fees)	Fixed costs based on coverage and capacity requirements
	CPE	Number of subscribers and deployment of OUs
5G FWA	Passive infrastructure (Shelter, tower)	Fixed costs based on coverage and capacity requirements
	Active infrastructure (e.g., baseband, gNodeB, backhaul, and software fees)	Fixed costs based on the spectrum band utilized, as well as capacity and coverage requirements
	CPE	Number of subscribers and deployment of OUs

DPU Distributed Point Unit; *OU* Outdoor Unit; *CPE* Customer Premises Equipment; *PON* Passive Optical Network; *ONT* Optical Network Terminal, *FSO* Free-Space Optics

a significant portion of both rural and urban areas [7, 39]. Table 1.2 presents the key cost drivers for each technology.

1.3 Fixed Wireless Broadband Overview

1.3.1.1 DSL Technology

DSL refers to a range of technologies that enable high-speed digital transmission over Unshielded Twisted Pairs (UTP) of copper wires. This twisted-wire-pair infrastructure was originally designed exclusively for POTS, intended to transmit a single voice signal within the frequency range of 300 Hz to 3.4 kHz [40, 41].

DSL technology has advanced with variations such as ADSL, VDSL, and G.Fast [42–45]. While VDSL and G.fast can provide speeds between 40–100 Mbps and even reach gigabit rates, respectively, actual xDSL broadband speeds in practice are frequently lower than 10 Mbps. This is due to factors such as the condition of the Fiber-to-the-Node (FTTN) system, the distance between the home and the exchange, and the condition of the copper lines. Additionally, the aging copper infrastructure contributes to high maintenance costs. Also, in many regions, particularly rural areas, xDSL struggles to achieve its optimal bandwidth due to signal degradation over long copper lines [2, 44].

G.fast employs a fiber connection to reach a distribution point, typically located within 250 m of the customer's premises, with the final segment delivered via copper. The performance experienced by the customer is influenced by the length of the copper sub-loop. Consequently, G.fast distributed point units (DPUs), whether reverse-powered or actively powered, have a maximum copper range defined by the performance targets of the network. The number of DPUs deployed is also affected by the number of premises served, with each DPU generally supporting up to 16 ports [46].

Cable companies and DSL service providers have competed fiercely to deliver triple play services, including Internet, Internet Protocol television (IPTV), and Voice over Internet Protocol (VoIP). As a result, DSL has become a leading broadband technology. To support multimedia services and data-intensive applications effectively, DSL providers are working to achieve higher data rates through various innovative technologies. One significant method for increasing speeds is bandwidth expansion. However, this approach can lead to crosstalk and interference within the system. In DSL networks, electromagnetic interference among lines within a cable bundle, caused by the electrical energy in each line, is a major barrier to performance improvement [47]. Dynamic Spectrum Management (DSM) techniques help mitigate this interference, enabling higher data rates. All DSL technologies operate at much higher frequencies, albeit with a reduced range [40, 42, 43].

1.3.1.2 Hybrid Fiber Coax

Significant effort has been invested in developing solutions to deliver broadband services to residential areas. Originally designed for one-way broadcast services, cable networks have evolved into a primary technology for providing two-way communication services to homes [48].

Modern cable operators now oversee large Hybrid Fiber-Coaxial (HFC) networks that connect numerous buildings. These networks utilize fiber optic system to deliver voice, video, and data traffic from central headends (i.e., data centers) to optical nodes within nearby neighborhoods. At these optical nodes, usually located within 500 m of the customer, the optical signal is converted into an radio frequency (RF) signal and then transmitted via sturdy, shielded coaxial cables to the customer premises. Conversely, RF signals traveling back to the headend are converted into the optical domain [9, 36].

A key component of modern HFC networks is the Data Over Cable Service Interface Specification (DOCSIS). DOCSIS is an industry standard created to provide high-quality video, audio, and interactive services, mainly Internet access, over HFC networks [2, 9]. A DOCSIS cable modem termination system (CMTS) is situated at the headend and connects to a cable modem (CM) at the customer's premises through the HFC network [49, 50].

CMs access upstream (US) channels employing a time division multiplexing method, which partitions the channel into mini-slots. Resource management agents at the CMTS oversee the allocation of these US mini-slots. According to DOCSIS, mini-slots can either be exploited using a contention-based approach with a contention resolution algorithm (CRA) or be allocated to particular stations. Typically, the method involves using individual mini-slots to convey bandwidth requests from CMs to the CMTS, thereby efficiently establishing a reservation channel [48].

The DOCSIS standard specifies the physical (PHY) and data link layer protocols and outlines the functionality required for providing two-way communication services to home users. It also outlines the mechanisms for resource allocation and quality assurance for the services provided [48, 49]. While HFC networks offer several strategic advantages for 5G densification, they face challenges such as latency and jitter. As a shared medium, HFC needs to manage traffic among multiple endpoints, which can lead to contention and latency spikes. Additionally, DOCSIS may face challenges in meeting the low latency and very high bandwidth demands of protocols such as the Common Public Radio Interface (CPRI) and Enhanced CPRI (eCPRI) [9, 49, 51–53].

The Low Latency xHaul (LLX) specification aims to overcome these issues by significantly minimizing jitter and latency for backhaul traffic over current HFC infrastructure. This improvement brings the latency of the entire traffic from the user equipment (UE) down to levels similar to those of fiber. The specification addresses various traffic flows, including IP Multimedia Subsystem (IMS) voice, signaling, video conferencing applications (like Cisco WebEx, Zoom, and Apple FaceTime), low-latency applications (such as mobile gaming), and ultra-Reliable Low Latency Communications (uRLLC) for 5G [8, 9, 54].

Like Fibre-to-the-Home (FTTH) Gigabit Passive Optical Network (GPON), the primary cost component for HFC networks involves civil works, which vary based on the fiber and coaxial cable topologies. Powered line amplifiers are employed to boost the signal, which weakens as the length of the coaxial cable increases. Taps and splitters are employed to distribute the signal to multiple premises. Each subscriber receives the necessary equipment, such as a cable modem, and is connected through in-building wiring [36, 49].

1.3.1.3 Fiber-Based Connections

Passive Optical Network (PON) is a fiber-optic system that operates as a passive broadband access solution. It transmits data US and downstream (DS) using different wavelengths and employs time-division multiplexing (TDM) for data transmission. A PON system comprises an Optical Line Terminal (OLT), an Optical Distribution Network (ODN), and Optical Network Units (ONUs), also known as Optical Network Terminals (ONTs). The OLT, situated at a Central Office (CO) or headend, connects to passive optical splitters in the field through feeder fibers. ODN connecting the OLT and ONUs utilizes optical fibers and passive components, minimizing electromagnetic interference, lightning effects, and providing strong environmental adaptability, as well as facilitating easy expansion and upgrades [55, 56]. Distribution fibers extend to multiple ONUs—typically up to 32—at customer premises, usually located within a 500 m radius of the splitter. In this context, PON employs a point-to-multipoint (P2MP) topology, allowing multiple ONUs to connect to a single PON port, thereby optimizing CO resources. Since only passive components are used throughout the network path, electrical power is required only at the endpoints. The optical link budget of a standard PON restricts the distance from the CO to the customer premises to about 20 km [8, 55, 57].

PON system has become extensively adopted because of its benefits, including high bandwidth, reliability, multi-service transmission capabilities, and cost-effectiveness. It also shares several advantages with HFC networks, such as widespread availability, ease of deployment, use of existing support structures and access agreements [44, 55].

Two standards organizations, The International Telecommunication Union Telecommunication Standardization Sector (ITU-T)/Full Service Access Network (FSAN) and IEEE, have been responsible for developing PON standards [56]. While the original standards accommodated gigabit speeds, both entities have subsequently introduced advanced PON specifications offering speeds reaching 10 Gbps and higher. Key distinctions between the IEEE and ITU-T PON standards include variations in line rates, optical budget coding gain, Forward Error Correction (FEC) types, packet segmentation (notably, Ethernet frames are not segmented in Ethernet PON, EPON), and other distinctions in the GPON Transmission Convergence (TC) layer compared to the EPON Media Access Control (MAC) layer [9, 55].

Additionally, several existing optical broadband network architectures effectively address performance bottlenecks by delivering services closer to customers. The overarching term for different fiber-optic broadband network architectures is fiber to the x (FTTx). Optical fiber can be used for either the entire or a portion of the last-mile network loop. The specific FTTx type—such as FTTN, FTTH, Fiber-to-the-Building (FTTB), Fiber-to-the-Premise (FTTP), or fiber to the cabinet (FTTC)—is primarily defined by how the optical fiber and copper lines connect to the CO and the customer premises equipment (CPE) [47, 55].

Both ITU-T and IEEE PON standards are employed to deliver FTTH and FTTP services. Also, they are extensively implemented in numerous urban areas. An FTTH network offers connection speeds of up to 100 Mbps and can be implemented using various architectural solutions, including P2P fiber, PON, and hybrid PON architectures [2, 47]. Recent analysis indicates that utilizing current FTTH networks can reduce 5G transport costs by over 50% in contrast to conventional methods like P2P dark fiber and microwave [8, 9, 55].

The primary costs associated with fiber-based connections involve civil work and obtaining government permits, which vary depending on whether the fiber cables are installed underground or above ground and if they are possessed or leased by communication service providers (CSPs) or wireless Internet service providers (WISPs). In some areas, deploying fiber solutions can be time-consuming, leading to opportunity costs [46].

ITU-T PON Standards

Following the release of the original GPON standard, the ITU-T has developed three additional PON standards. These standards, as well as their respective US and DS data rates, are detailed in Table 1.3. The standard facilitates a seamless enhancement path to higher-capacity PONs by employing regulated system components, such as coexistence elements. The wavelength plan for US and DS also supports the deployment of 10 Gbps P2P wavelengths (P2P WDM) within the unified ODN as the PON. Research is ongoing for access network bitrates of 25 Gbps and higher.

Historically, PON solutions were unable to handle traditional CPRI fronthaul (i.e., option 8 split) due to the stream-based nature of legacy CPRI interfaces, which grow in proportion to the number of radio transmit/receive chains and require consistently high bandwidth, making them suitable only for P2P dark fiber system [51, 58]. However, with the introduction of the Third-Generation Partnership Project (3GPP) Release 16 and 17 services, including Massive Machine Type Communications (mMTC) and uRLLC, there is a need for a more flexible topology driven by the virtualization of Radio Access Network (RAN) functions [59]. MNOs are transitioning towards extensively scalable and programmable cloud-native architectures, which involve disaggregation of network functions to facilitate more adaptable service deployment. Consequently, network functions must effectively communicate using various service types over a shared transport network infrastructure. Flexible service reconfiguration and service-level agreement (SLA) guarantees will be essential to realize this concept [9, 60].

Furthermore, it is important to note that the functional split among the Radio Unit (RU), Distributed Unit (DU), and Central Unit (CU) in a disaggregated RAN can significantly increase the need for low latency and high bandwidth compared to traditional all-in-one gNB backhaul solutions [60]. PON technology is versatile enough to support various function splits, including low-level split (LLS) fronthaul, high-level split (HLS) midhaul, and backhaul, thereby addressing requirements related to jitter, latency, scalability, and connection bandwidth [47, 60]. Chapter 3 provides comprehensive information on how the 5G FWA transport network can utilize functional split options developed for mobile networks to achieve the necessary flexibility and adapt to future demands.

1.3 Fixed Wireless Broadband Overview

Table 1.3 PON standards evolution

		Alternative	Standard	Data rate (Gbps)		
				DS	US	References
ITU-T		GPON	G.984x	2.5	1	[9, 56]
		XG-PON	G.987	10	2.5	[9, 56]
		NG-PON2	G.989	4×10	4×10	[9]
		XGS-PON	G.9807.1	10	10	[9, 56]
		G.HSP	G.9804	50	12.5/25/50	[56, 61, 62]
IEEE		EPON	802.3ah	1	1	[9, 56]
		10G EPON	802.3av	10	10	[9, 56]
		NG-EPON	802.3ca	25/50	25/50	[9, 56, 62, 62, 63]

HSP Higher Speed PON

IEEE PON Standards

Similar to the ITU-T, the IEEE has been diligently developing and advancing PON standards starting with the introduction of the first one in 2004. The latest standards include 10G EPON and Next-generation EPON (NG-EPON). The IEEE PON standards along with their respective DS and US data rates are outlined in Table 1.3. In an EPON, transmission from the OLT to the ONUs (DS transmission) occurs in a P2MP configuration. To deliver Ethernet frames to the ONUs, the OLT broadcasts them. The frames are then accurately routed to the ONUs based on their MAC addresses. Nonetheless, because ONUs are unable to directly exchange messages with one another, they need to utilize the entire trunk fiber resources when sending data to the OLT (US transmission). To optimize the use of trunk fiber resources, EPONs employ various multiple access techniques, including TDM, wavelength-division multiplexing (WDM), code-division multiple access (CDMA), and combinations of these methods. However, the existing EPON standards fail to fully accommodate the future needs of EPON applications [64].

On June 4, 2020, the IEEE 802.3 Working Group standardized NG-EPON, also referred to 25G/50G-EPON. It offers an economical and feasible solution for future implementations. One of the goals of NG-EPONs is to boost the bitrate of a single channel to 25 Gb/s while maintaining a cost per bit comparable to the existing 10G PONs. Another goal of NG-EPONs is to multiplex multiple 25 Gb/s wavelength channels to deliver aggregated data rates of $N \times 25$ Gb/s for each ONU. As a result, NG-EPONs utilize channel bonding technology, which allows multiple wavelength channels to be bonded to an ONU simultaneously, thereby achieving higher peak transmission rates. Consequently, distributing traffic across multiple wavelength channels offers enhanced scheduling flexibility, increased peak US rates, and improved US transmission performance (such as reduced latency). However, this approach requires expensive ONUs equipped with multiple transceivers, which further raises the overall cost of NG-EPONs. Thus, choosing an appropriate wavelength allocation strategy is crucial when planning future NG-EPONs. In this context, $1 \times 25G$ and $2 \times 25G$ NG-EPON variants have been defined to reach peak aggregate rates [8, 64–66].

1.3.1.4 Ethernet

As MNOs expand their networks to deliver higher capacity and speeds using 5G, they must also focus on optimizing the total cost of ownership (TCO) for their RAN. To achieve this, many service MNOs are adopting cloud RAN (C-RAN) frameworks to consolidate baseband processing tasks. This approach enhances the efficiency of baseband processing resources and enhances operational efficiency at cell sites [52, 67, 68].

Additionally, C-RAN allows MNOs to tailor their network to various applications by strategically placing storage and compute resources throughout the network. For instance, cloud edge solutions can be positioned closer to cell sites to support applications requiring low latency. However, shifting resource-intensive functions to a centralized site necessitates adhering to stringent transport latency and bandwidth demands typical of fronthaul networks. Currently, 4G fronthaul networks often use semi-proprietary protocols like CPRI, transmitted over dark fiber. Although these methods fulfill the required transport specifications, they are expensive to deploy and maintain. Additionally, they limit the capacity to support several services [53, 69].

Recent advancements in radio encapsulation and time-sensitive networking (TSN) techniques have made it feasible to deploy fronthaul networks using Ethernet solutions. Ethernet, which has become the standard for cell site backhaul, is preferred for its cost-effectiveness, flexibility in services, compatibility across multiple vendors, and widespread use. Its integration into fronthaul networks promises significant cost reductions for C-RAN architectures and allows operators to benefit from the most recent advancements in network orchestration and automation. Additionally, this shift enables extensive deployment of robust monitoring and control, and opens the opportunity of consolidating the entire 5G xHaul networks and other fixed services into a unified transport network [37, 53, 54, 67].

1.3.2 Fixed Wireless Access Network

FWA technology, which has been in use for over a decade across 3G and 4G networks, primarily serves rural areas lacking fixed broadband or experiencing low speeds. It offers significant advantages, including typically lower initial deployment expenditures compared to fixed alternatives and capital efficiency by leveraging existing spectrum holdings [70]. For CSPs, FWA presents an attractive opportunity to deliver network bandwidth comparable to fiber optics in certain areas [71]. This technology enables CSPs to deliver high-speed, low-latency broadband to suburban and rural consumers, where the cost of fiber installation and maintenance has been prohibitive. Additionally, FWA allows for rapid deployment and a swift return on investment (ROI) [16, 72].

1.3 Fixed Wireless Broadband Overview

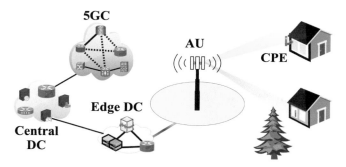

Fig. 1.2 5G-based FWA deployment

As depicted in Fig. 1.2, installing Access Units (AUs), the 5G base stations (BSs), on building rooftops or utility poles near existing fiber backhaul can provide a neighborhood or district with high-capacity, high-density wireless network coverage. Once a subscription is activated, users can easily set up a 5G CPE, much like a cable modem or DSL, by placing it in a window close to the AU for instant connectivity. This setup offers a broadband experience comparable to FTTH, without the need for a contractor to excavate the front lawn to install a physical fiber connection to the home [71, 73].

The potential of widespread FWA continues to expand with each contemporary generation of wireless technology, and 5G is no exemption. Currently, 5G is emerging as a key technology for FWA in both small and medium-sized enterprises (SMEs) and residential settings. As discussed in Chap. 2, 5G is projected to deliver 10 to 100 times the capacity of 4G, enabling the deployment of cost-effective FWA schemes on a larger scale. This enhanced connectivity can improve the economic viability of FWA and equip operators with the capabilities to meet market demands that were previously considered economically unfeasible [2, 39].

Moreover, 5G FWA can support the transmission of signals through both Wi-Fi 5 and Wi-Fi 6 technologies, enabling convenient Internet connectivity for multiple devices. It also addresses IoT use cases that are not necessarily mobile, such as surveillance and security cameras. To fully exploit the market potential of FWA, operators and the mobile industry should ensure interoperability and scalability for cost-effective FWA by adopting 3GPP specifications [25, 47].

Even now, with LTE utilizing 40 MHz of bandwidth, FWA frequently offers a viable business solution as an enhancement to MBB. This potential is expected to strengthen further as LTE continues to evolve. As explained in Chap. 4, the transition to 5G promises to elevate FWA significantly, owing to 5G's unprecedented technology options, including access to larger spectrum segments and benefits like low latency and substantial capacity improvements. This makes 5G FWA a strong alternative to FTTH [2, 39]. As illustrated in Fig. 1.3, FWA provides a chance to maximize the benefits of 5G deployment by simultaneously handling the two primary 5G use cases—MBB and fixed wireless. During the day, 5G

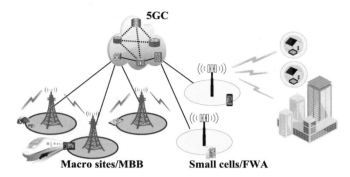

Fig. 1.3 5G for MBB and fixed wireless

beams cater to mobile users outdoors, and in the evening, they may be rerouted to an FWA device when users return home. This dual-purpose functionality enhances the justification for 5G deployment, presenting it as a sustainable and cost-effective technology [39].

Additionally, 5G-based FWA is anticipated to deliver highly attractive and robust services with sustainable high rates sufficient to meet future home use needs. Also, it is expected to provide maximum cell rates that surpass those of most high-capacity fixed technologies, which would otherwise require substantial investments in deep-fiber fixed access infrastructure.

5G FWA could also enhance current fixed BBA in dense urban areas by achieving greater peak rates to satisfy growing latency and bandwidth demands without extensive physical infrastructure upgrades. Notably, 5G FWA seems equipped to address bandwidth saturation issues resulting from high demand for residential services like IPTV. Furthermore, the extremely low latency of 5G access could be a crucial factor for enabling future applications.

1.4 5G FWA Opportunities

A 5G fixed wireless solution represents an ideal choice for MNOs aiming to offer HBB services in both rural and urban areas. Broadband solutions based on FWA are particularly advantageous in sparsely populated rural regions, where traditional fixed-line options are economically challenging to implement. It is crucial to highlight that coverage and overall performance are heavily influenced by the frequency range utilized, the operating environment, and the placement of the terminal antenna [39]. In this context, the innovative wireless technologies and capabilities of 5G NR can address the challenges of connecting rural homes by utilizing coverage extension features. Tools such as Integrated Access and Backhaul (IAB) can enhance network density at a relatively low cost, delivering high-speed connections to multiple homes through massive MIMO and beamforming technologies [7, 74]. Thus, 5G operators offering services based on FWA can provide numerous advantages to both urban and rural consumer segments, as discussed in the following subsections.

1.4 5G FWA Opportunities

1.4.1 Improved Broadband Speed

FWA leveraging 3GPP radio access technology (RAT) utilizes radio advancements like the carrier aggregation (CA), LTE air interface, multi-antenna technologies, and innovative modulation schemes. In contrast to mobile phones, devices used in wireless transport networks are not limited by strict power supply constraints, enabling enhanced performance. Additionally, multi-antenna CPEs and high-gain outdoor CPEs are offered as alternatives. These benefits enable FWA to achieve fiber-like performance, though they also lead to increased implementation costs [25, 75].

4G LTE technologies, incorporating the aforementioned radio innovations, currently deliver peak rates of 1 Gb/s. This performance surpasses copper wire transmission and matches that of fiber networks. Through the use of CA and multi-antenna technologies, throughput is significantly enhanced. Specifically, 4×4 MIMO schemes combined with CA of five 20 MHz carriers substantially increase throughput. In contrast, 5G utilizes even wider bandwidth and massive MIMO schemes, aiming for peak user rates of up to 20 Gb/s within up to 1 GHz bandwidth. This capability allows 5G to support a greater number of users per BS while delivering high user rates [25]. Consequently, 5G-based FWA has the potential to elevate broadband speeds to meet or exceed targets, offering a significant boost in user experience for consumers in rural areas [7, 76].

Additionally, the broadband value proposition extends beyond merely offering higher data speeds. It encompasses the ability to access multimedia services and other applications that require dependable connectivity, including e-learning. This feature facilitates the delivery of quad-play packages, a globally successful business model. With FWA, operators can offer connectivity while also bundling innovative services like VoIP and 4K or 8K resolution Ultra High Definition (UHD) TV to fascinate consumers [25].

1.4.2 Fast Network Development

The full benefits of FWA are realized when deployment and operational costs remain comparatively low relative to wireline infrastructure for delivering broadband connections to users [25]. For instance, FTTH requires extensive time to secure RoW permissions and involves significant civil work (ducts or poles), which prolongs the deployment process, making it time-consuming and expensive. The process typically spans some weeks between service request to activation, and in the most challenging scenario, fiber connections may be inaccessible for several months. FWA changes this scenario entirely. 5G FWA offers a more flexible and rapid deployment option, as it does not require laying cables or RoW permissions. Also, it can be set up relatively quickly, reducing visual clutter from utility poles and underground cables, and cutting service activation period to as little as 1 or 2 d [76].

Additionally, a significant portion of the cost savings is due to lower civil engineering expenses, which are among the most intricate and expensive aspects of network implementation and operation. This complexity arises from the need for licenses, trench digging, pole installation, tower erection, and other related tasks [25]. As a result, the average cost per connection decreases, allowing networks with the same TCO to serve a greater number of households than FTTH. This enables operators to build their user base much more rapidly than with FTTH [2, 76].

Moreover, incorporating radio technologies like massive MIMO with 3D beamforming can lower costs to 80% of those associated with wireline connections, once the initial investment in these technologies is recouped. This cost reduction allows operators to deploy data connections more rapidly and flexibly for their intended applications [25].

1.4.3 Homogenized Wider Coverage

Wired technologies often show significant differences in speed between urban and rural areas. For instance, in certain regions, average download speeds for cable and fiber can be four to ten times greater in urban areas than rural ones. However, 3GPP RAT can offer coverage comparable to fiber networks [25]. 5G networks, designed to cover larger geographic areas than previous generations, enable high-speed Internet access in more remote or rural locations. This eliminates the need to wait for fiber optic upgrades from your ISP. Consequently, a well-implemented 5G-based FWA solution can deliver a uniform user experience in both rural and urban areas [7].

1.4.4 Flexibility

There are various user segments with distinct needs: some are price-sensitive and seek lower-cost data traffic, while others prioritize high service quality. FWA utilizes 4G/5G wireless technologies, enabling network sharing. This approach allows MNOs to offer both prepaid and postpaid plans tailored to diverse customer groups with varying service needs [2].

1.4.5 Close the Digital Divide

In rural areas, there is often a lack of broadband options compared to urban areas. To address this issue, LTE operators with nationwide coverage can offer viable broadband solutions to these underserved regions by quickly and efficiently deploying 5G on their existing infrastructure. This expansion allows rural consumers to access a greater variety of broadband providers, helping to bridge the digital divide in many locations [76].

1.5 FWA Technical Challenges and Solutions

Although 5G offers substantial opportunities, deploying a high-speed FWA network across various scenarios presents several technical challenges. This section outlines these technical hurdles and proposes innovative solutions to address them, with the goal of simplifying the implementation process for providers. By tackling these challenges, the section seeks to enhance the economics of deploying FWA broadband services and provide advanced ultra-broadband capabilities to consumers and businesses with limited or no existing service options.

1.5.1 Spectrum-Based

The main challenges with FWA primarily stem from the spectrum utilized. To offer a viable alternative to current wireline broadband, wireless solutions must meet capacity and throughput expectations comparable to FTTH services. Given the scarcity of spectrum below 3 GHz, Millimeter Wave (mmWave) is emerging as the favored choice for FWA solutions. Consequently, this necessitates the introduction of innovative tools and technologies to address the limitations of high-frequency radio signals during planning and deployment [73].

Furthermore, service providers need to expand their coverage through small cell schemes, but this will result in higher infrastructure costs for each household served. To address this, advanced antenna technologies like massive MIMO and beamforming, along with sophisticated signal processing, can direct the radio signal into a focused, flashlight-like beam that targets each device with precision. This approach helps concentrate the radio energy in a smaller area, significantly enhancing the range and penetration of mmWave signals [73]. However, these technologies are both complex and costly.

1.5.2 Deployment-Based

To guarantee seamless connectivity, it is essential to address factors such as signal propagation, tower placement, and coverage areas. Challenges related to placement include deployments with high-density tree foliage, non-ideal Line-of-Sight (LoS), and low-emissivity glass (a kind of glass specifically engineered to minimize thermal radiation). These issues typically result in significantly reduced signal quality, resulting in subpar throughput and user experience [73, 77, 78]. For instance, the height of the transmitter and receiver antennas is crucial for achieving LoS. This ensures that the link maintains first Fresnel zone clearance between the them. The required antenna height for achieving LoS varies and depends on the terrain and the height of any obstructions along the link [72]. The radius of the nth Fresnel zone circle can be represented as [79, 80]

$$r_n = \sqrt{\frac{n\lambda d_2}{d_1 + d_2}} \tag{1.1}$$

where r_n is the number of radius of Fresnel zone, λ represents the wavelength of the transmitted signal in meters, d_1 denotes the distance between the transmitter and the obstacles, and d_2 is the distance between the obstacle and the receiver. The wireless link is regarded as free space if 80% of the first Fresnel zone is unobstructed by obstacles [80].

Furthermore, a more focused and powerful signal can penetrate foliage more effectively and enables strategic beam manipulation, such as reflecting it off surfaces to connect with equipment that do not have a direct LoS with the AU. This capability can be realized through the implementation of beamforming technology [74]. To address scenarios where adequate coverage cannot be delivered to an in-home (indoor) 5G CPE—such as when the only available window does not face the AU or when buildings or street fixtures obstruct the optimal beam path—there are alternative form factors available, such as outdoor 5G CPEs. These rugged, all-weather outdoor CPEs can be adaptably installed on building rooftops or fitted on exterior facades, providing an optimal deployment location. They require just a short cable run to connect to indoor Wi-Fi or Ethernet routing equipment [73].

Additionally, a Machine Learning (ML)-based RF planning software tool can identify environmental factors such as building geometry, trees, and construction materials. It then utilizes ray tracing techniques to quickly model the interaction of radio beams with the environment and connect with prospective equipment. This tool can replace conventional drive testing approaches, where network planners must drive through neighborhoods using costly radio testing devices, a time-consuming process that can span several days for a single area [73]. Chapter 6 presents AI-based enhancement methods aimed at improving the effectiveness of wireless network coverage planning tools.

1.5.3 Interoperability

Globally interoperable devices and network infrastructure, along with a broad community that collaboratively develops standards and pushes technological boundaries, are considered fundamental to the success of mobile communications. Although FWA deployment figures will be significantly lower than those for IoT and cellular, attaining scale is still a crucial objective for realizing its full potential [25].

Currently, mobile operators are evaluating a range of technology solutions. However, without a unified technology standard from the global mobile operator community, the economic benefits of network deployment solutions and devices will not be fully realized, and technological fragmentation must be minimized to achieve economies of scale. Embracing and ensuring alignment with RATs defined by 3GPP offers the greatest opportunity for promoting scalability and interoperability [25].

1.5.4 Network Congestion

As the need for fast-speed Internet increases, FWA providers encounter major difficulties in handling network congestion and maintaining high levels of Quality of Experience (QoE). Advanced traffic management solutions offer a comprehensive solution to tackling these problems. By employing real-time congestion management, dynamic application recognition, and enforcement of fair use policy, FWA providers can lower costs, improve customer satisfaction, and expand their networks more effectively [81].

1.6 FWA Implementation Considerations

The increased access to fast-speed broadband through 4G LTE and 5G NR presents MNOs with the opportunity to provide broadband services to households and SMEs using FWA. This is particularly beneficial in suburban and rural areas, where deploying fiber and copper connections can be cost-prohibitive, and spectrum resources are commonly underutilized. Furthermore, as legacy infrastructure like DSL becomes outdated, FWA offers an appealing, high-speed, and cost-effective alternative [7].

Most residential use cases involve subscribers living in detached buildings, often referred to as single-family homes, while about 25% live in multi-dwelling units (MDUs). An MDU refers to any building housing multiple households, such as townhouse, apartment complexes, or high-density residential buildings. The proportion of houses in MDUs is significantly greater than in single-family homes in densely populated urban areas [7]. Therefore, 5G operators looking to provide FWA services to residential customers across various types of housing and demographics must consider critical factors such as spectrum, site locations, capacity, product speeds, and CPE, as illustrated in Fig. 1.4.

1.6.1 Spectrum

Spectrum is undoubtedly a crucial factor influencing an operator's 5G FWA strategy. The availability of additional spectrum in mid and high bands, along with the advancement of 5G NR in low bands, offers significant expansion opportunities for FWA [7]. Figure 1.5 offers a qualitative assessment of various NR bands, evaluating their performance with respect to latency, capacity, and coverage.

By combining mid-band and high-band frequencies via Dual Connectivity (DC) or CA, MNOs can deliver a high level of service to users. This approach offers exceptionally high throughput to customers near the cell through high-band carriers while also providing excellent service to users located at greater distances through mid-band spectrum [7]. Additionally, a complete 5G network utilizing CA facilitates the development of a fully coordinated, multi-layer 5G network, employing various layers with distinct characteristics and advan-

Fig. 1.4 Factors for MNOs to consider about FWA

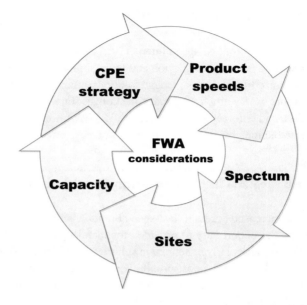

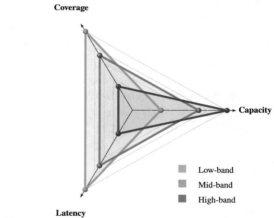

Fig. 1.5 Comparison of different key features of 5G spectrum bands

tages, as illustrated in Fig. 1.6. In such a network, 5G carriers across different bands—low, mid, and high—can be aggregated, ensuring optimal performance and flexibility to achieve service differentiation [75]. This topic is examined more thoroughly in Chap. 7.

1.6.1.1 4G Spectrum

Technically, there are no restrictions on the frequency bands that can be used for deploying FWA. In LTE network implementations, MNOs typically favour using 800 MHz, 1.8, and 2.1 GHz bands for rural and suburban areas, while opting for 2.3 and 2.6 GHz bands in

1.6 FWA Implementation Considerations

Fig. 1.6 A fully integrated 5G network featuring carrier aggregation

urban areas. This approach can be similarly applied to FWA. Nonetheless, because of the potential to counteract the drawbacks of increased propagation loss, considering higher carrier frequencies, like those within the 3.5 GHz range, is also viable, as larger bandwidths are more easily accessible there [25].

The majority of current FWA deployments utilize medium- and high-frequency bands. However, in areas with low demand for mobile broadband, deploying FWA using low and mid-frequency bands is feasible without adversely affecting the user experience [25]. Dedicated carriers for FWA in medium- and high-frequency bands are typically above 2 GHz, with common choices being 2.3 GHz (band 40), 2.6 GHz (band 41), or 3.5 GHz (band 42 or band 43). By employing multiple antenna systems at both ends of the link and using higher transmit power than the standard 23 dBm used for mobile device, operators can offset the increased propagation loss associated with higher frequency bands. The available bandwidth allows for optical fiber-class broadband performance and facilitates the delivery of high-quality FWA. Based on this, innovative services like real-time and multimedia applications can be supported [25].

1.6.1.2 5G Spectrum

Initial 5G deployments utilize a combination of different spectrum bands. The 3.5 GHz band is commonly used for early cellular communications, while different sets of mmWave spectrum are being adopted by operators now and will continue to be in a later phase, based on accessibility and their strategic plans [25].

While sub-6 GHz spectrum provides much enhanced coverage than mmWave, its narrower bandwidth and limited availability have driven several MNOs to initially roll out 5G FWA in mmWave bands, where beamforming and massive MIMO technologies can be utilized more effectively. The primary concern with mmWave lies in its higher penetration loss through structures and the limited distance that BSs can support. In mmWave spectrum, radio link performance significantly improves with a direct LoS between the transmitter and the receiver, necessitating the installation of many external antennas, which complicates deployment and operations. This complexity decreases the cost-effectiveness of a wireless last-mile connection. This is due to the additional engineering visits and installations that may be needed. Ericsson's experiments confirm that 3.5 GHz provides substantially improved coverage compared to mmWave in various scenarios. For instance, a BS operating at 3.5 GHz can offer a minimum range of 1,800 m to an indoor terminal antenna, outperforming all instances of 28 GHz except in cases where the BS is positioned above clutter and the terminal antenna is mounted on a rooftop [25, 39, 74].

1.6.2 Sites

The concept of FWA relies on the viability of delivering last-mile connections via wireless means. The density of BSs and physical location are crucial for 5G MNOs aiming to provide FWA. Often, BSs are strategically placed to offer last-mile connectivity to households where wired options, like fiber, are too costly, providing MNOs an edge over traditional wireline providers. Established mobile operators also benefit from the ability to cost-effectively upgrade their existing LTE sites to 5G technology. 5G deployment can be strategically planned to prioritize areas with both mobile and fixed wireless demand, creating synergies between these two services. Operators can utilize both street pole-mounted small cells and macro sites to provide FWA. Though macro sites typically offer advantages due to their greater height and transmission power, providing coverage over several miles compared to the few hundred feet of small cells. Additionally, elevated macro sites, particularly in rural areas, can efficiently deliver last-mile connectivity through fixed wireless to households in low-density areas [7].

1.6.3 Capacity

Consumer usage patterns, representing the total needs of the entire subscribing households within a coverage area, are crucial for managing network capacity by FWA operators. 5G FWA operators can leverage *oversubscription* as a factor for network planning and scaling to optimize user experience during peak times. The concept of oversubscription is based on the observation that not every household is actively using the service simultaneously even during busy hours. The majority of telecommunications networks leverage this multiplexing gain to allocate their capacity lower than the peak demand level [7].

Moreover, the traffic patterns for home and MBB differ, which can benefit wireless operators. Traditionally, HBB usage peaks on weekends compared to weekdays. However, with the rise of remote work and online learning, weekday usage has significantly increased. Conversely, smartphone traffic has historically been greater during the week due to activities

1.6 FWA Implementation Considerations

related to mobility, but this pattern has also shifted as a result of changes brought about by the pandemic [7].

The criterion for determining capacity requirements involves ensuring that broadband services are provided to all subscribing homes concurrently, while maintaining a basic level of quality during peak traffic periods. This standard applies separately to both Downlink (DL) and Uplink (UL) scenarios, respectively as [47, 82]

$$C_{DL}^{PC} \geq N_{Sub}^{PC} \times R_{DL}^{MinQ}, \tag{1.2a}$$

$$C_{UL}^{PC} \geq N_{Sub}^{PC} \times R_{UL}^{MinQ} \tag{1.2b}$$

where C_{DL}^{PC} and C_{UL}^{PC} denote the system's DL and UL throughput per cell, respectively; N_{Sub}^{PC} represents the count of subscribing households in every cell; and R_{DL}^{MinQ} and R_{UL}^{MinQ} indicate the demanded throughput per household to achieve the minimum quality of Internet service, $MinQ$, for the DL and UL, respectively.

MNOs that have predominantly focused on mobile user traffic often possess surplus capacity in various parts of their network that can be leveraged for FWA. Based on the placement of BS sites, operators can utilize this underused capacity to accommodate both fixed and mobile applications using their current spectrum assets [7].

Beyond the surplus capacity, there is also the potential to make use of spectrum that has yet to be activated. While spectrum licenses cover uniform geographical areas, mobile usage does not always align with this, leading to some spectrum bands either being entirely unused or underutilized. 5G operators can relatively easily deploy equipment to activate additional spectrum bands on current sites to accommodate FWA subscribers [7].

Established MNOs must effectively handle the allocation of resources among fixed and mobile consumers and expedite the transfer of spectrum assets from 4G LTE to 5G NR. This transition is crucial to leverage the enhancements offered by 5G, such as improved security, network slicing, and spectrum efficiency. Operators with dedicated spectrum for 5G are well-positioned to meet the increasing demands from 5G applications, including home Internet [7, 83].

1.6.4 Service Speeds

Product or service speeds denote the peak advertised DL and UL data rates that a provider offers on a *best-effort* basis. These data rates can fluctuate based on the technology employed and the geographical area covered, with the Federal Communications Commission (FCC) mandating baseline speeds of 25 Mbps for DL and 3 Mbps for UL. In urban areas, where demand and competition are higher, broadband offerings are increasingly targeting upper speed tiers. 5G operators providing FWA can leverage advancements in NR standards to enhance spectral efficiency, and in turn, throughput and cell capacity. As mentioned in Chap. 2, the 5G NR standard significantly improves spectral efficiency relative to LTE by increas-

Table 1.4 Comparisons of broadband speeds

	Technology	Download speed (average)	References
DSL	ADSL/ADSL2+	24 Mbps	[46, 47, 70]
	FTTC/VDSL2	200 Mbps	[46, 70]
	G.Fast	100 Mbps–1 Gbps	[46, 70]
Fibre	FTTP/H	2.4–40 Gbps	[46, 47, 70]
Cable	DOCSIS 3.1	10 Gbps	[46, 47, 70]
Satellite	LEO	50–500 Mbps	[46, 70]
	GEO	12–50 Mbps	[46]
FWA	4G (LTE)	Up to 100 Mbps	[46, 70]
	5G	1–10 Gbps*	[46, 70, 82]

*Depends on the density of cell sites and the spectrum band employed

ing the throughput per unit of bandwidth (Hz). This improvement is due to advancements in engineering, signaling, and CPE technology. Key factors comprise antenna gains, CPE power, Massive MIMO, Sound Reference Signal (SRS) antenna selection (which allows CPE to swap antennas for optimal signal reception quality), and strategic CPE positioning (e.g., near windows, walls, or rooftop mounts) [7].

Table 1.4 shows a comparison of download speeds across different broadband technologies, highlighting that 5G FWA can offer up to 10 times or more speed improvement over 4G, depending on the deployment scenario. The capacity for mid-band and mmWave spectrum will also increase, and combined with enhancements in CPE technology, this will lead to greater spectral efficiency and increased revenue per Hz [70].

The criterion for determining system speed centers on meeting the specified DL and UL speed demands. Specifically, the FWA method must be able to provide the specified DL and UL speeds simultaneously to at least $X\%$ of the subscribed households. This principle applies separately to both DL and UL speeds, respectively as [47, 82]

$$\frac{C_{DL}^{PC}/S_{DL}^{PC}}{N_{Sub}^{PC}} \times 100\% \geq X\%, \tag{1.3a}$$

$$\frac{C_{UL}^{PC}/S_{UL}^{PC}}{N_{Sub}^{PC}} \times 100\% \geq X\% \tag{1.3b}$$

where S_{DL}^{PC} and S_{UL}^{PC} denote the system's UL and DL speeds, respectively.

1.6.4.1 Guaranteed Household Percentage

The Guaranteed Household Percentage (GHP) represents the proportion of subscribing households guaranteed to achieve the S_{DL}^{PC}/S_{UL}^{PC} download/upload speeds simultaneously,

1.6 FWA Implementation Considerations

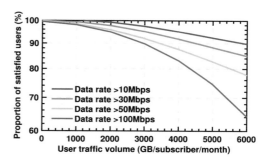

Fig. 1.7 Proportion of satisfied users as a function of traffic

while ensuring that all other households receive the minimum QoS. Given that the system is designed to provide $Z\%$ of subscribed households with S_{DL}^{PC}/S_{UL}^{PC} speeds concurrently, the DL and UL throughputs of the system can be defined as [47, 82]

$$C_{DL}^{PC} = R_{DL}^{QoS} \times N_{Sub}^{PC} (1 - Z\%) + S_{DL}^{PC} \times N_{Sub} (Z\%) \text{ Mbps}, \quad (1.4a)$$

$$C_{UL}^{PC} = R_{UL}^{QoS} \times N_{Sub}^{PC} (1 - Z\%) + S_{UL}^{PC} \times N_{Sub} (Z\%) \text{ Mbps} \quad (1.4b)$$

where R_{DL}^{QoS} and R_{UL}^{QoS} denote the DL and UL speeds required to provide minimal QoS to all subscribing households simultaneously.

The system design, built on the 5G NR concept operating at 28 GHz with a 200 MHz bandwidth and employing beamforming and multiuser MIMO (MU-MIMO), can be used to explore the connection between user traffic volume and the proportion of satisfied users [39, 74, 84]. This analysis can be conducted through polynomial regression. In this context, by denoting $P(V, R)$ as the proportion of satisfied users given user traffic volume V (GB/subscriber/month) and data rate R (Mbps), the proportion of satisfied users as a function of traffic for data rates can be defined by the third-degree polynomial given as

$$P(V) = aV^3 + bV^2 + cV + d \quad (1.5)$$

where a, b, c, and d are the coefficients of the polynomial fit, which can be found by solving a set of linear equations obtained through the least squares method.

As illustrated in Fig. 1.7, most users are able to achieve the required or target data rates based on the traffic load. However, as traffic load rises, user throughputs tend to decline due to increased queuing and heightened interference. For a target data rate of 20 Mbps, approximately 95% of users can attain a rate that exceeds this target [39, 74, 84].

1.6.5 Devices

When providing FWA, it is essential to consider various device form factors as well as traditional routers and CPE. In the context of mmWave 5G, devices must be capable of operating effectively in outdoor environments. Since the radio link can significantly degrade due to obstacles or the lack of LoS caused by glass materials, the devices need to be durable and able to withstand all weather conditions. This durability can enhance network deployment economics by decreasing the number of BSs needed for adequate coverage [25]. In this regard, the receiver within the FWA system is a crucial factor influencing both service quality and coverage boundaries. Significant recent advancements in FWA include the performance improvements offered by the 5G standard and innovations in CPE that facilitate easier onboarding [70]. This topic is explored in greater detail in Chap. 5.

1.7 Conclusion

FWA has become a highly appealing and cost-effective option to fixed broadband for various applications, especially in rural and suburban areas where the expense of fiber deployment poses a barrier. FWA advances the Broadband for All objective and contributes to bridging the Digital Divide. This hybrid approach integrates elements of both shared mobile networks (i.e., 4G and 5G) and dedicated fixed networks to provide Internet services to end users. Instead of using fiber, cable, or copper for last-mile transmission, FWA leverages cellular spectrum to connect homes. Additionally, 5G technologies can address challenges related to broadband status and speed.

As 5G evolves to support multi-access networks, FWA can complement existing fixed broadband access by enhancing peak rates for home CPEs. With ongoing advancements such as increased spectrum allocation, beamforming, improved terminals, optimized media distribution, and the virtualization of RAN and core networks, FWA is emerging as a more viable option to traditional fixed broadband access.

References

1. A. Ericsson, L. Falconetti, H. Olofsson, J. Edstam, T. Dahlberg, Closing the digital divide with mmWave extended range for FWA. Ericsson Technol. Rev. **2022**(11), 2–11 (2022)
2. Wireless Fiber, Huawei Technologies, 4G/5G FWA Broadband Industry White Paper (2019). Accessed 04 Jul 2024
3. M. Lyon, K. Hafner, *Where Wizards Stay Up Late: The Origins of The Internet*. Simon & Schuster (1999)
4. J. Naughton, *A Brief History of the Future: The Origins of the Internet* (A Phoenix paperback, Phoenix, 2000)
5. J. Abbate, *Inventing the Internet* (MIT Press, Inside Technology, 2000)
6. Strategies and policies for the deployment of broadband in developing countries, International Telecommunication Union, Development Sector, Report Study Group 1 Question 1 (2021), https://www.itu.int/dms_pub/itu-d/opb/stg/D-STG-SG01.01.2-2021-PDF-E.pdf. Accessed 23 Apr 2024

7. Fixed Wireless Access with 5G Networks, 5G Americas, White Paper (2021), https://www.5gamericas.org/wp-content/uploads/2021/11/5G-FWA-WP.pdf. Accessed 21 Apr 2024
8. E. Ganesan, A.T. Liem, I.-S. Hwang, QoS-aware multicast for crowdsourced 360° live streaming in SDN aided NG-EPON. IEEE Access **10**, 9935–9949 (2022)
9. Innovations in 5G Backhaul Technologies: IAB, HFC & Fiber, 5G Americas, White Paper (2020), https://www.5gamericas.org/wp-content/uploads/2020/06/Innovations-in-5G-Backhaul-Technologies-WP-PDF.pdf. Accessed 15 Jul 2024
10. N. Gonzales, *Rural Americans vs. Urban Americans* (Public Affairs Council, Report, 2023), https://pac.org/impact/rural-americans-vs-urban-americans. Accessed 15 May 2024
11. L. Dijkstra, E. Papadimitriou, *Annex: Using a New Global Urban-Rural Definition, Called the Degree of Urbanisation, to Assess Happiness* (The World Happiness Report, Report, 2020), https://happiness-report.s3.amazonaws.com/2020/WHR20_ChAnnex.pdf. Accessed 15 May 2024
12. B.K. Behera, R. Prasad, Shyambhavee, Chapter 4—Health-care information technology and rural community, in *Healthcare Strategies and Planning for Social Inclusion and Development*, ed. by B.K. Behera, R. Prasad, Shyambhavee (Academic Press, 2022), pp. 85–117, https://www.sciencedirect.com/science/article/pii/B9780323904469000046
13. P.J. Smailes, N. Argent, T.L. Griffin, Rural population density: its impact on social and demographic aspects of rural communities. J. Rural Stud. **18**(4), 385–404 (2002), https://www.sciencedirect.com/science/article/pii/S0743016702000335
14. E.J. Oughton, Policy options for broadband infrastructure strategies: a simulation model for affordable universal broadband in Africa. Telematics Inf. **76**, 101908 (2023), https://www.sciencedirect.com/science/article/pii/S0736585322001411
15. Y. Chu, H. Ahmadi, D. Grace, D. Burns, Deep learning assisted fixed wireless access network coverage planning. IEEE Access **9**, 124 530–124 540 (2021)
16. Bridging the digital divide: Extended-range millimeter-wave 5G Fixed Wireless Access, Ericsson and US cellular, Case Study (2022), https://www.ericsson.com/4a9aa7/assets/local/cases/customer-cases/2022/uscellular-bridging-digital-divide.pdf. Accessed 20 Apr 2024
17. K.R. Chaudhuri, E. Cunha Neto, L. Falconetti, R. Fassbinder, S. Guirguis, A. Halder, M. Irizarry, R.D. Patel, N. Saxena, S. Sorlescu, Extended range mmWave for fixed wireless applications, in *2021 97th ARFTG Microwave Measurement Conference (ARFTG)* (2021), pp. 1–4
18. P. Tedeschi, S. Sciancalepore, R. Di Pietro, Satellite-based communications security: a survey of threats, solutions, and research challenges. Comput. Netw. **216**, 109246 (2022), https://www.sciencedirect.com/science/article/pii/S138912862200319X
19. S. Yuan, M. Peng, Y. Sun, X. Liu, Software defined intelligent satellite-terrestrial integrated networks: insights and challenges. Digit. Commun. Netw. **9**(6), 1331–1339 (2023), https://www.sciencedirect.com/science/article/pii/S2352864822001316
20. M. Kang, S. Park, Y. Lee, A survey on satellite communication system security. Sensors **24**(9) (2024), https://www.mdpi.com/1424-8220/24/9/2897
21. R. Gallardo, B. Whitacre, The Real Digital Divide? Advertised Vs. Actual Internet Speeds, Updater, News (2020), https://pcrd.purdue.edu/the-real-digital-divide-advertised-vs-actual-internet-speeds/. Accessed 22 Jul 2024
22. J. Supan, Advertised vs. actual internet speeds, Are you getting the internet speeds you pay for? Allconnect, Internet (2024), https://www.allconnect.com/blog/advertised-vs-actual-internet-speeds. Accessed 22 Jul 2024
23. J. Weinerman, Internet Speeds: Advertised vs Actual, Updater, Internet (2022), https://updater.com/guides/internet-speeds-advertised-vs-actual. Accessed 22 Jul 2024

24. Declaratory Ruling, Order, Report and Order, and Order on Reconsideration, Federal Communications Commission, Order FCC 24-52 (2024), https://docs.fcc.gov/public/attachments/FCC-24-52A1.pdf. Accessed 22 Jul 2024
25. Fixed Wireless Access: Economic Potential and Best Practices, GSMA, White Paper (2018), https://www.gsma.com/solutions-and-impact/technologies/networks/wp-content/uploads/2018/08/Fixed-Wireless-Access-economic-potential-and-best-practices.pdf. Accessed 05 Jul 2024
26. What Is 5G NR? 5G New Radio Standard Explained, Celona, Position Paper (2022), https://www.celona.io/5g-lan/5g-nr. Accessed 23 Apr 2024
27. I.A. Alimi, P.P. Monteiro, Performance analysis of 5G and beyond mixed THz/FSO relaying communication systems. Opt. Laser Technol. **176**, 110917 (2024), https://www.sciencedirect.com/science/article/pii/S003039922400375X
28. D. Montolio, F. Trillas, Regulatory federalism and industrial policy in broadband telecommunications. Inf. Econ. Policy **25**(1), 18–31 (2013), https://www.sciencedirect.com/science/article/pii/S016762451300005X
29. A. Picot, C. Wernick, The role of government in broadband access. Telecommun. Policy **31**(10), 660–674 (2007), https://www.sciencedirect.com/science/article/pii/S0308596107000833
30. J.S. Park, Wide Area Networks, in *Encyclopedia of Information Systems*, ed. H. Bidgoli (Elsevier, New York, 2003), pp. 649–660, https://www.sciencedirect.com/science/article/pii/B0122272404001970
31. O. Font-Bach, N. Bartzoudis, A. Pascual-Iserte, D.L. Bueno, A real-time MIMO-OFDM mobile WiMAX receiver: architecture, design and FPGA implementation. Comput. Netw. **55**(16), 3634–3647 (2011), mobile WiMAX, https://www.sciencedirect.com/science/article/pii/S1389128611000910
32. OECD broadband statistics update, Organisation for Economic Co-operation and Development, Press Release (2024), https://www.oecd.org/en/about/news/press-releases/2024/03/broadband-statistics-update.html. Accessed 20 May 2024
33. Network Densification, HUBER+SUHNER, White Paper (2021), https://www.hubersuhner.com/en/markets/communication/mobile-network/small-celldensification. Accessed 25 Jul 2024
34. Small cell densification, CommScope, White Paper (2022), https://www.commscope.com/solutions/5g-mobile/small-cell-densification/. Accessed 25 Jul 2024
35. K. Mun, Small cells or massive MIMO: the best way to densify (Analyst Angle). RCR Wirel. Insights (2022), https://www.rcrwireless.com/20220523/analyst-angle/small-cells-or-massive-mimo-the-best-way-to-densify-analyst-angle. Accessed 25 Jul 2024
36. M. Garcia, D. Garcia, V. Garcia, R. Bonis, Analysis and modeling of traffic on a hybrid fiber-coax network. IEEE J. Select. Areas Commun. **22**(9), 1718–1730 (2004)
37. I.A. Alimi, A.L. Teixeira, P.P. Monteiro, Toward an efficient C-RAN optical fronthaul for the future networks: a tutorial on technologies, requirements, challenges, and solutions. IEEE Commun. Surv. Tutor. **20**(1), 708–769 (2018)
38. I. Alimi, A. Shahpari, A. Sousa, R. Ferreira, P. Monteiro, A. Teixeira, Challenges and opportunities of optical wireless communication technologies, in *Optical Communication Technology*, ed. by P. Pinho (IntechOpen, Rijeka, 2017). https://doi.org/10.5772/intechopen.69113
39. K. Laraqui, S. Tombaz, A. Furuskär, B. Skubic, A. Nazari, E. Trojer, Fixed wireless access on a massive scale with 5G. Ericsson, Technol. Rev. **94** (2017), https://www.ericsson.com/assets/local/publications/ericsson-technology-review/docs/2017/2017-01-volume-94-etr-magazine.pdf
40. O. Naparstek, K. Cohen, A. Leshem, Parametric spectrum shaping for downstream spectrum management of digital subscriber lines. IEEE Commun. Lett. **16**(3), 417–419 (2012)

41. W. Foubert, C. Neus, L. Van Biesen, Y. Rolain, Exploiting the phantom-mode signal in DSL applications. IEEE Trans. Instrum. Meas. **61**(4), 896–902 (2012)
42. C. Leung, S. Huberman, K. Ho-Van, T. Le-Ngoc, Vectored DSL: potential, implementation issues and challenges. IEEE Commun. Surv. Tutor. **15**(4), 1907–1923 (2013)
43. P. Tsiaflakis, M. Diehl, M. Moonen, Distributed spectrum management algorithms for multiuser DSL networks. IEEE Trans. Signal Process. **56**(10), 4825–4843 (2008)
44. N. Keukeleire, B. Hesmans, O. Bonaventure, Increasing broadband reach with hybrid access networks. IEEE Commun. Stand. Mag. **4**(1), 43–49 (2020)
45. F. Mazzenga, R. Giuliano, F. Vatalaro, Sharing of copper pairs for improving DSL performance in FTTx access networks. IEEE Access **7**, 6637–6649 (2019)
46. F. Agnoletto, A. Goel, P. Castells, The 5G FWA opportunity: disrupting the broadband market, GSMA Intelligence, White Paper (2021), https://data.gsmaintelligence.com/api-web/v2/research-filedownload?id=66289674&file=141021-5G-FWA-Opportunity.pdf. Accessed 05 Jul 2024
47. I.A. Alimi, R.K. Patel, N.J. Muga, A.N. Pinto, A.L. Teixeira, P.P. Monteiro, Towards enhanced mobile broadband communications: a tutorial on enabling technologies, design considerations, and prospects of 5G and beyond fixed wireless access networks. Appl. Sci. **11**(21) (2021), https://www.mdpi.com/2076-3417/11/21/10427
48. C. Cesar Heyaime-Duverge, V.K. Prabhu, Statistical multiplexing of upstream transmissions in DOCSIS cable networks. IEEE Trans. Broadcast. **56**(3), 296–310 (2010)
49. J. Wang, Z. Jia, L.A. Campos, L. Cheng, C. Knittle, G.-K. Chang, Delta-sigma digitization and optical coherent transmission of DOCSIS 3.1 signals in hybrid fiber coax networks. J. Lightwave Technol. **36**(2), 568–579 (2018)
50. C. Cui, S. Park, RoIP compression method in burst noise upstream environment for CATV network. IEEE Trans. Broadcast. **67**(1), 351–355 (2021)
51. M. Klinkowski, M. Jaworski, Dedicated path protection with wavelength aggregation in 5g packet-optical Xhaul access networks. J. Lightwave Technol. **41**(6), 1591–1602 (2023)
52. I.A. Alimi, P.P. Monteiro, A.L. Teixeira, Analysis of multiuser mixed RF/FSO relay networks for performance improvements in Cloud Computing-Based Radio Access Networks (CC-RANs). Opt. Commun. **402**, 653–661 (2017), https://www.sciencedirect.com/science/article/pii/S0030401817305734
53. I.A. Alimi, R.K. Patel, A. Zaouga, N.J. Muga, A.N. Pinto, A.L. Teixeira, P.P. Monteiro, *6G CloudNet: Towards a Distributed, Autonomous, and Federated AI-Enabled Cloud and Edge Computing* (Springer International Publishing, Cham, 2021), pp. 251–283. https://doi.org/10.1007/978-3-030-72777-2_13
54. J. Prados-Garzon, T. Taleb, M. Bagaa, Optimization of flow allocation in asynchronous deterministic 5G transport networks by leveraging data analytics. IEEE Trans. Mob. Comput. **22**(3), 1672–1687 (2023)
55. A.F. Pakpahan, I.-S. Hwang, Flexible access network multi-tenancy using NFV/SDN in TWDM-PON. IEEE Access **11**, 42 937–42 948 (2023)
56. White Paper on 50G PON Technology, ZTE, White Paper V2.0 (2018). Accessed 15 Jul 2024
57. M. Nakamura, H. Ueda, S. Makino, T. Yokotani, K. Oshima, Proposal of networking by PON technologies for full and ethernet services in FTTx. J. Lightwave Technol. **22**(11), 2631–2640 (2004)
58. P. Öhlén, B. Skubic, A. Rostami, M. Fiorani, P. Monti, Z. Ghebretensaé, J. Mårtensson, K. Wang, L. Wosinska, Data plane and control architectures for 5G transport networks. J. Lightwave Technol. **34**(6), 1501–1508 (2016)
59. I. Chih-Lin, H. Li, J. Korhonen, J. Huang, L. Han, RAN revolution with NGFI (xhaul) for 5G. J. Lightwave Technol. **36**(2), 541–550 (2018)

60. I.A. Alimi, P.P. Monteiro, Functional split perspectives: a disruptive approach to RAN performance improvement. Wirel. Pers. Commun. **106**(1), 205–218 (2019). https://doi.org/10.1007/s11277-019-06272-7
61. HSP: Higher Speed Passive Optical Networks, ITU-T Study Group 15, White Paper G.9804 (2021). Accessed 15 Jul 2024
62. P. Torres-Ferrera, H. Wang, V. Ferrero, M. Valvo, R. Gaudino, Optimization of band-limited DSP-aided 25 and 50 Gb/s PON using 10G-class DML and APD. J. Lightwave Technol. **38**(3), 608–618 (2020)
63. V. Houtsma, D. van Veen, Bi-directional 25G/50G TDM-PON with extended power budget using 25G APD and coherent detection. J. Lightwave Technol. **36**(1), 122–127 (2018)
64. C. Zhang, M. Yang, W. Zheng, Y. Zheng, Y. Wu, Y. Zhang, Analysis of wavelength deployment schemes in terms of optical network unit cost and upstream transmission performance in NG-EPONs. J. Opt. Commun. Netw. **13**(9), 214–223 (2021)
65. IEEE Standard for Ethernet Amendment 9: Physical Layer Specifications and Management Parameters for 25 Gb/s and 50 Gb/s Passive Optical Networks, in *IEEE Std 802.3ca-2020 (Amendment to IEEE Std 802.3-2018 as amended by IEEE 802.3cb-2018, IEEE 802.3bt-2018, IEEE 802.3cd-2018, IEEE 802.3cn-2019, IEEE 802.3cg-2019, IEEE 802.3cq-2020, IEEE 802.3cm-2020, and IEEE 802.3ch-2020)* (2020), pp. 1–267
66. IEEE Draft Standard for Ethernet Amendment: Physical Layer Specifications and Management Parameters for 25 Gb/s and 50 Gb/s Passive Optical Networks, in *IEEE P802.3ca/D3.1* (2020), pp. 1–34
67. I.A. Alimi, A.M. Abdalla, A. Olapade Mufutau, F. Pereira Guiomar, I. Otung, J. Rodriguez, P. Pereira Monteiro, A.L. Teixeira, *Energy Efficiency in the Cloud Radio Access Network (C-RAN) for 5G Mobile Networks* (John Wiley & Sons, Ltd., 2019), pp. 225–248, https://onlinelibrary.wiley.com/doi/abs/10.1002/9781119491590.ch11
68. I.A. Alimi, P.P. Monteiro, A.L.J. Teixeira, Outage probability of multiuser mixed RF/FSO relay schemes for heterogeneous cloud radio access networks (H-CRANs). Wirel. Pers. Commun. **95**, 27–41 (2017). https://doi.org/10.1007/s11277-017-4413-y
69. I.A. Alimi, R.K. Patel, A. Zaouga, N.J. Muga, Q. Xin, A.N. Pinto, P.P. Monteiro, Trends in cloud computing paradigms: fundamental issues, recent advances, and research directions toward 6G fog networks, in *Moving Broadband Mobile Communications Forward*, ed. by A. Haidine (IntechOpen, Rijeka, 2021). https://doi.org/10.5772/intechopen.98315
70. T. Hatt, P. Jarich, E. Kolta, J. Joiner, 5G fixed wireless: a renewed playbook, GSMA Intelligence, White Paper (2021), https://data.gsmaintelligence.com/api-web/v2/research-file-download?id=60620869&file=020321-5G-fixed-wireless.pdf. Accessed 21 May 2024
71. A.P.K. Reddy, M.S. Kumari, V. Dhanwani, A.K. Bachkaniwala, N. Kumar, K. Vasudevan, S. Selvaganapathy, S.K. Devar, P. Rathod, V.B. James, 5G new radio key performance indicators evaluation for IMT-2020 radio interface technology. IEEE Access **9**, 112 290–112 311 (2021)
72. N. Saba, J. Salo, K. Ruttik, R. Jäntti, Using existing base station sites for 5G millimeter-wave fixed wireless access: antenna height and coverage analysis. IEEE Wirel. Commun. Netw. Conf. (WCNC) 1–6 (2024)
73. 5G Fixed Wireless Access, Samsung, White Paper (2018), https://images.samsung.com/is/content/samsung/p5/global/business/networks/insights/white-paper/samsung-5g-fwa/white-paper_samsung-5g-fixed-wireless-access.pdf. Accessed 05 Jul 2024
74. M. Girnyk, H. Jidhage, S. Faxér, Broad beamforming technology in 5G massive MIMO. Ericsson Technol. Rev. **2023**(10), 2–6 (2023)
75. What, Why and How: the Power of 5G Carrier Aggregation. Ericsson, Tech. Rep. (2021), https://www.ericsson.com/en/blog/2021/6/what-why-how-5g-carrier-aggregation. Accessed 22 Apr 2024

76. K. Aldubaikhy, W. Wu, N. Zhang, N. Cheng, X. Shen, mmWave IEEE 802.11ay for 5G fixed wireless access. IEEE Wirel. Commun. **27**(2), 88–95 (2020)
77. J. Zhang, C. Masouros, Learning-based predictive transmitter-receiver beam alignment in millimeter wave fixed wireless access links. IEEE Trans. Signal Process. **69**, 3268–3282 (2021)
78. Z. El Khaled, W. Ajib, H. Mcheick, Log distance path loss model: application and improvement for sub 5 GHz rural fixed wireless networks. IEEE Access **10**, 52 020–52 029 (2022)
79. A.D.S. Braga, H.A.O.D. Cruz, L.E.C. Eras, J.P.L. Araújo, M.C.A. Neto, D.K.N. Silva, G.P.S. Cavalcante, Radio propagation models based on machine learning using geometric parameters for a mixed city-river path. IEEE Access **8**, 146 395–146 407 (2020)
80. J. Joo, D.S. Han, H.-J. Jeong, First Fresnel zone analysis in vehicle-to-vehicle communications, in *International Conference on Connected Vehicles and Expo (ICCVE)* (2015), pp. 196–197
81. M. Goldshtein, Bandwidth on the move: innovative approaches to FWA challenges, Mobile Europe, White Paper (2024), https://www.mobileeurope.co.uk/bandwidth-on-the-move-innovative-approaches-to-fwa-challenges/. Accessed 10 Jul 2024
82. L. Zhang, Y. Wu, S. Lafleche, X. Huang, S. Dumoulin, R. Paiement, A. Florea, Capability evaluation of fixed wireless access systems to deliver broadband internet services (Communications Research Centre Canada, CRC Technical Report, 2020), https://www.ic.gc.ca/eic/site/139.nsf/eng/00014.html
83. H.-S. Lee, S. Moon, D.-Y. Kim, J.-W. Lee, Packet-based fronthauling in 5G networks: network slicing-aware packetization. IEEE Commun. Stand. Mag. **7**(2), 56–63 (2023)
84. C. Ranaweera, P. Monti, B. Skubic, E. Wong, M. Furdek, L. Wosinska, C.M. Machuca, A. Nirmalathas, C. Lim, Optical transport network design for 5G fixed wireless access. J. Lightwave Technol. **37**(16), 3893–3901 (2019)

Network Architecture and Evolution

2.1 Introduction

The surge in mobile traffic, driven by the expansion of various mobile and multimedia services, has led to widespread use of terrestrial telecommunications networks, which now cater to most of the world's population. A crucial factor impacting the user experience is the last-mile technology, which is often the primary speed constraint in wireless networks. This bottleneck results in high Operating Expenses (OPEX) and Capital Expenditure (CAPEX) for network operators. As outlined in Chap. 1, traditional last-mile solutions, such as broadband over copper cables (e.g., xDSL, including ADSL, ADSL+, and VDSL, alternatively referred to as FTTC), depend on current landline telephone systems and are limited in speed. In recent years, optical access systems for both mobile and fixed services have undergone substantial advancements in infrastructure and management [1–5].

At present, the prevalent last-mile technologies are FTTP and FTTC, which employ optical fiber to provide Internet services directly to consumers' locations. FTTP offers faster speeds and significantly higher bandwidth compared to conventional copper wire. Nevertheless, implementing FTTP demands substantial investment in additional infrastructure. Even in nations with well-developed infrastructure, fiber connections are not universally accessible. As a result, it may be considered cost-prohibitive for network operators in sparsely populated rural areas [4, 6, 7].

FWA has become a crucial technology for bridging the digital divide and addressing gaps in fiber networks. It is rapidly becoming a key application for 5G [8]. One of the core strengths of 5G is its design as a future-proof technology that delivers broadband services cost-effectively. FWA utilizing 5G NR technology connects the 5G system to CPE at specific locations, such as homes or businesses, providing high-speed Internet access [8, 9]. With the deployment of 5G, existing FWA equipment can reach speeds of up to 150 Mbps on 5G NR bands under 7 GHz. Expanding 5G NR into the mmWave spectrum will enable FWA to reach multi-Gbps speeds, matching those of high-speed fiber networks [6, 10].

By leveraging mmWave spectrum, 5G FWA offers a scalable and competitive alternative that supports high-speed Internet for various applications, including high-definition streaming, gaming, augmented reality (AR) and virtual reality (VR) [10, 11]. To guarantee an optimal user experience in a 5G network, future network infrastructure must enhance overall traffic capacity, peak data rates, synchronization, latency, automation, security, and novel interfaces. For instance, advancements like novel coding techniques, massive MIMO antenna technology, and extensive channel bandwidths provided by mmWave spectrum have increased peak data rates ranging from 1 to 10 Gbps and higher. Additionally, the new frame structure of 5G NR has decreased latency from over 10 ms to under 1 ms, compared to earlier 4G technology [12–15].

2.2 Wireless Next-Generation Technologies

The unfulfilled need for broadband access can be cost-effectively addressed with FWA, especially by leveraging the widespread infrastructure and global reach of the 3GPP mobile technologies like High-Speed Downlink Packet Access (HSDPA), 4G LTE, and 5G NR. MNOs can leverage the ongoing advancements in 3GPP technologies by implementing FWA alongside MBB across both current and new spectrum bands, due to the effective spectrum-sharing options available between the two services. Additionally, the robust 3GPP framework for network equipment and device chipsets, built on MBB, underpins a solid and future-proof foundation for the evolution of FWA [8, 16–18].

2.2.1 5G New Radio

5G signifies the fifth generation of wireless communication technology designed for digital cellular networks. It is poised to be the most advanced solution for meeting the demands of an increasingly sophisticated mobile market. 5G aims to elevate mobile networks beyond just connecting people, to also link and control objects, machines, and devices. It establishes novel benchmarks for exceptional performance and efficiency, facilitating enhanced user experiences, smart homes, and connections to emerging industries. Looking ahead, 5G will be closely incorporated with artificial intelligence (AI), big data, cloud computing, AR, and VR. It will enhance connectivity across various domains and serve as a crucial foundation for the digital transformation of every industry. The diverse variety of 5G use cases will provide a strong foundation for entrepreneurial growth and revolution [19, 20].

Furthermore, 5G represents the latest generation of mobile technology, offering low latency, ultra-fast speeds, and exceptional reliability. The 5G NR interface is designed to meet the advanced demands of diverse usage scenarios, addressing significant growth in connectivity and traffic density/volume [19]. It provides an adaptable air interface to meet key use cases for 5G communications: enhanced MBB (eMBB), uRLLC, and massive IoT

2.2 Wireless Next-Generation Technologies

support [21–24]. The benefits of 5G NR include reduced latency, increased user capacity, network slicing, and improved speed. These advancements enable technologies like AR, smart city IoT applications, remote-controlled vehicles, and more [25, 26]. The NR air interface supports deployments in both sub-6 GHz bands and newly available spectrum in mmWave and cmWave bands [20, 27].

A key element of the NR air interface is its compatibility with large-scale antenna arrays and massive MIMO. Massive MIMO primarily aims to improve coverage and capacity, which are essential for addressing the growing need for data services. With data service demand expected to increase by around 50% annually in the foreseeable future, the advanced features of NR, along with additional spectrum allocations, will be crucial for meeting these rising data demands [27].

5G NR depends significantly on MIMO technology. MIMO, in conjunction with beamforming, facilitates several crucial capabilities provided by 5G NR. Beamforming enables multi-antenna BSs to electronically direct spatial beams toward intended receivers, improving reliability and coverage. Additionally, beamforming enhances system capacity by allowing multiple spatially distinct users to be multiplexed to share the same time-frequency resources [5, 28, 29].

The transition from 4G LTE to 5G NR represent a significant advancement, aimed at enhancing mobile telecommunications and unlocking a wide range of new prospects. In addition to substantially improving key performance metrics, the transition introduces a more versatile and powerful radio architecture. This includes the addition of mmWave frequency spectrum alongside existing 4G LTE-Advanced (LTE-A) and novel sub-6 GHz frequency bands [30]. A notable change from the LTE era is the anticipated introduction of lower-cost routers (i.e., CPE) with advanced performance capabilities, leveraging massive MIMO and multiple antenna technologies, among other innovations [31–33].

Plans are also in place to extend 5G frequency bands to encompass both licensed and unlicensed mmWave spectrum. Additionally, 5G NR supports both TDD and frequency-division duplex (FDD) operations, offering wider channel bandwidths, advanced multi-antenna architectures, increased maximum power for UE, and higher-order modulation schemes [23, 34]. Designers of 5G NR RF front-ends (RFFE) will gain valuable insights by grasping these trends along with the latest RF hardware and technologies required to meet emerging challenges [24, 30].

2.2.2 5G Use Cases

5G inherently supports a range of emerging use cases beyond eMBB, including uRLLC and mMTC, as outlined by the ITU-R 5G-IMT-2020 Vision [4, 12, 19, 23]. Additionally, 3GPP defines FWA as a specific use case of eMBB [35]. These technologies are already being utilized within 5G NR. Consequently, these new use cases are imposing tighter demands on the transport networks that underpin the 5G infrastructure.

2.2.2.1 Ultra-Reliable Low-Latency Communication

5G NR supports uRLLC, delivering packet transmission latencies as low as a few milliseconds and network reliability exceeding five nines. These performance benchmarks make uRLLC highly attractive to enterprises and service providers with stringent network requirements. Examples of uRLLC applications include smart city infrastructure, AR, autonomous robotics and vehicles, industrial IoT, and automation [23, 25].

2.2.2.2 Massive Machine-Type Communications

5G NR supports extensive mMTC across a wide range of applications and devices by enabling more effective signal processing and reducing energy consumption. It will be crucial for the widespread deployment of 5G IoT, as businesses expand their networks with additional sensors and applications [23, 25, 36]. Examples of mMTC applications include industrial sensors that track safety systems such as pipe pressure and air quality; healthcare sensors for tracking patients, inventory, and life-saving equipment; general enterprise device management; as well as sensors that oversee machine performance, health, and product tracking across the assembly process.

2.2.2.3 Enhanced Mobile Broadband

5G NR will significantly boost MBB performance and influence the future advancements in cellular technology. Its higher efficiency, coupled with reduced energy consumption and increased data rates, benefits both customers and businesses. By leveraging higher frequency bands within the mmWave spectrum, 5G NR can achieve faster user connections by integrating with existing 4G infrastructure through dynamic spectrum sharing. This approach facilitates the broad adoption of 5G NR and makes deployment more cost-effective for network operators. The benefits and applications of eMBB include lower energy consumption across cellular networks, faster download speeds, more efficient video streaming, improved service for commercial carriers, and enhanced coverage across smart cities [22, 25].

2.3 RAN Architecture

The RAN architecture is a crucial element in wireless communication systems, serving as the essential connection linking the core network and user devices. The RAN manages all radio-related functions, enabling the transmission and reception of data through the air interface, and is instrumental in shaping the overall performance, efficiency, and capabilities of mobile networks.

The evolution of RAN architecture is key to the progression of mobile networks, from the early 2G and 3G systems to the advanced 4G and next-generation 5G networks. Modern RAN architectures, which adopt centralized, open, and virtualized approaches, are designed to address the increasing demands for high performance, flexibility, and cost-efficiency

2.3 RAN Architecture

in the rapidly advancing field of wireless communication. As we transition into the 5G period, a diverse array of new services and technologies will be swiftly rolled out. These include Network Function Virtualization (NFV), mmWave spectrum, Containerized Network Functions, uRLLC, mMTC, Network Slicing, Multi-Access Edge Computing (MEC), Vertical Services, and multi-radio dual connectivity (MR-DC) Protocol architectures such as Evolved-Universal Terrestrial Radio Access (E-UTRA)-NR (EN-DC) and NR Dual Connectivity (NR-DC), among others [4, 26, 37–39]. To effectively support these innovations, various approaches to distributing functions within the RAN have been proposed.

2.3.1 3GPP RAN Architecture

The allocation of RAN functions across the radio antenna site and central locations is crucial for determining transport requirements. These functions encompass RF signal processing and various layers of the protocol stack (PS), comprising the PHY, radio link control (RLC), medium access control (MAC), radio resource control (RRC), and packet data convergence protocol (PDCP) layers [12, 39–42]. Figure 2.1a illustrates the interaction between these RAN functions, UE, and the 5G Core (5GC) network.

The 3GPP has established a next-generation RAN (NG-RAN) architecture, which divides the functionality of the 5G NR BS (gNB) into two conceptual units: a CU and a DU. In the 3GPP framework, the CU interfaces with the 5GC network through the NG interface, while it connects to the DU by the F1 interface, as depicted in Fig. 2.1b [12, 43]. The specific split between the centralized and distributed components varies depending on service demands and network scenarios [37, 44].

The 3GPP has explored several functional splits among the CU and DU, considering a total of eight possible options. These include five HLS options and three LLS options, as illustrated in Fig. 2.2 [4, 12]. For HLS, the 3GPP Release 15 work item is concentrating on standardizing Option 2 (PDCP/high RLC split). In the LLS category, the prominent contenders are Option 6 (MAC/PHY split) and Option 7 (intra-PHY split), with further alternatives like 7-1, 7-2, and 7-3. These possible options are evaluated in [45–47]. The following subsections will delve into and elaborate on several options currently being considered by 3GPP.

Fig. 2.1 Functions distribution in **a** RAN and **b** 3GPP NG-RAN architecture

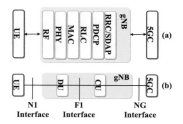

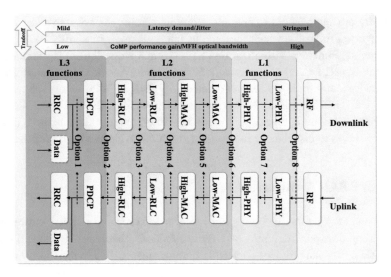

Fig. 2.2 Function split options in the disaggregated RAN. MAC: Medium Access Control, RF: Radio Frequency, PDCP: Packet Data Convergence Protocol, RRC: Radio Resource Control, RLC: Radio Link Control, PHY: Physical

2.3.1.1 Option 8 Split

This option provides an ideal scheme for effective traffic management and adaptable hardware resource allocation. It imposes no limitations on the type of centralized processing that can be efficiently realized. Additionally, it simplifies BS design by eliminating the need for local processing, aside from the necessary mobile fronthaul (MFH) protocol and digital filtering. While this approach significantly simplifies the DU and enhances the efficiency of cell sites [48, 49], it also imposes strict latency requirements on the MFH due to centralized implementations of frequency, power control, and time synchronization [50]. As a result, this option centralizes high-layer processing functions, which affects bandwidth demands and imposes strict latency constraints on the MFH [4]. Consequently, the CPRI data rate for multiple sectors and antennas configurations can be specified as [4, 45, 51–53]

$$R_{D_{CPRI}} = N_s \, M \, R_s \, N_{res} \, v C_w \, C, \qquad (2.1)$$

where C_w denotes the factor for the CPRI codewords, N_{res} represents the sample width (bits/sample), R_s is the sampling rate employed for digitization (samples/second/carrier), N_s and M are the number of sectors and the number of antennas per sector, respectively, C denotes the coding factor, and a factor $v = 2$ is used to account for the intricate nature of the IQ samples.

2.3 RAN Architecture

2.3.1.2 Low Layer Split

Since Option 8 is a proven choice and Options 6 and 7 are acknowledged as feasible LLS applications according to 3GPP terminology [54], the placement of the split within LLS introduces significant disparities in transport requirements. Consequently, the chosen split location can substantially impact both the transport network scheme and the overall RAN architecture.

Option 7

The PHY functions are divided among the DU and CU in Option 7, providing further benefits for resource sharing and load balancing. This PHY function split can be implemented in various fashions among the units, with each sub-option presenting diverse functionalities and bandwidth requirements [55]. The broadly recognized sub-option splits are Options 7-1, 7-2, and 7-3 [36]. These sub-option splits are applicable to DL transmission, while only Options 7-1 and 7-2 are applicable for UL transmission. Notably, the necessary bandwidth for Option 7-3 in UL is significantly higher than in DL, rendering its UL application less attractive because of the need for transmitting soft bits for FEC decoding [48]. The MFH bandwidth demanded for Option 7-2 increases with the number of streams and the system bandwidth, while for Options 7-1 and 8, it scales with the number of antenna ports and the RF system bandwidth. This dependency on antenna ports results in the MFH bandwidth requirements for these options being significantly higher than for Option 7-2. Typically, the demanded bandwidth for DL and UL in this option is expressed, respectively, as [4, 36, 56–58]

$$R_{DL}^{\text{Intra-PHY}} = 2 N_{\text{res}}^{DL} N_{sc} N_{\text{symb}} L_s^{\text{MIMO}_{DL}} N_{\text{max}}^{UE} + \text{MAC}_{\text{info}}^{DL^{\text{subopt}}}, \quad (2.2a)$$

$$R_{UL}^{\text{Intra-PHY}} = 2 N_{\text{res}}^{UL} N_{sc} N_{\text{symb}} L_s^{\text{MIMO}_{UL}} N_{\text{max}}^{UE} + \text{MAC}_{\text{info}}^{UL^{\text{subopt}}}, \quad (2.2b)$$

where N_{max}^{UE} denotes the maximum number of UEs, N_{symb} is the number of symbols in a transmission time interval (TTI), N_{sc} represents the number of subcarriers in the resource block, $\text{MAC}_{\text{info}}^{(\cdot)}$ denotes the MAC information for each sub-option (UL or DL) [58], and $L_s^{\text{MIMO}_{DL}}$ and $L_s^{\text{MIMO}_{UL}}$ are the MIMO layer scaling factors for the DL and UL, respectively. The parameter $L_s^{\text{MIMO}(\cdot)}$ is defined as [4, 36, 56, 57]

$$L_s^{\text{MIMO}(\cdot)} = L_n^{\text{base}(\cdot)} / L_n^{\text{LTE}(\cdot)}, \quad (2.3)$$

where $L_n^{\text{base}(\cdot)}$ and $L_n^{\text{LTE}(\cdot)}$ are the baseline and LTE reference parameters, respectively.

Typically, when the MAC is located in the CU, intra-PHY sub-option splits efficiently support a range of features, including CA, network MIMO, DL/UL Coordinated Multi-Point (CoMP), and joint processing (JP) [4, 40]. Likewise, a PHY split can accommodate additional features without changes to the radio equipment (RE), as it holds almost all baseband functionalities [54]. This significantly simplifies the DU and, consequently, the cell sites, allowing them to be placed on streetlamp or utility poles [4, 48].

Moreover, when sufficient number of low-layer functions are centralized, the primary advantage of LLS is the improved pooling gains and coordination between adjacent cells. Conversely, when low-layer functions are more decentralized, the primary benefit is the substantially reduced transport requirement in comparison to Option 8 split, which enables easier scalability for massive MIMO applications. However, in comparison to HLS, intra-PHY sub-options require lower latency and higher capacity for MFH [59]. This could necessitate additional resources to support the network, potentially increasing energy consumption and costs of the system [4, 48].

Option 6

Option 6 entails implementing all Layer 1 (L1) processing locally within the DU, with L2 and L3 functions handled by the CU [50]. In contrast to Option 8, which generally transmits IQ data, Option 6 transmits MAC frame data, leading to a substantial reduction in MFH bandwidth. Consequently, MFH bandwidth is directly dependent on the real user throughput. Additionally, this option offers pooling gains compared to HLS options and supports advanced radio coordination techniques due to centralized scheduling [40, 50].

Moreover, Option 6 simplifies DU architecture, making it not only more cost-effective but also simpler to install and maintain. This reduction in DU footprint facilitates easier installation on utility poles or street lamp poles [4, 48].

Although the MFH bandwidth is nearly reduced to match the wireless data rate, implementing centralized MIMO processing is comparatively challenging due to the need for the DU to handle computationally intensive PHY layer functions [60]. Consequently, the DU is comparatively complex [48]. The needed bandwidth for the UL and DL in this option can be defined as [4, 36, 56–58]

$$R_{DL}^{\text{MAC-PHY}} = \left(R_p^{DL} + S_o^{S\&C}\right) BW_s^{SY} L_s^{\text{MIMO}_{DL}} M_s^{DL}, \quad (2.4a)$$

$$R_{UL}^{\text{MAC-PHY}} = \left(R_p^{UL} + S_o^{R}\right) BW_s^{SY} L_s^{\text{MIMO}_{UL}} M_s^{UL}, \quad (2.4b)$$

where R_p^{DL} and R_p^{UL} are the reference LTE peak data rates for the DL and UL, respectively, S_o^R denotes the signal overhead because of the UL-PHY's response to the schedule, $S_o^{S\&C}$ is the signal overhead owing to scheduling/control signaling to DL-PHY, BW_s^{SY} represents the system bandwidth scaling, and M_s^{DL} and M_s^{UL} represent the modulation order (QAM) scaling for the DL and UL, respectively. Parameters $M_s^{(\cdot)}$ and BW_s^{SY} can be defined as [36, 56, 57]

$$M_s^{(\cdot)} = M_s^{\text{base}(\cdot)} / M_s^{\text{LTE}(\cdot)}, \quad (2.5a)$$

$$BW_s^{SY} = B^{\text{base}} / B^{\text{LTE}}, \quad (2.5b)$$

where B^{base} and $M_s^{\text{base}(\cdot)}$ are the baseline parameters, and B^{LTE} and $M_s^{\text{LTE}(\cdot)}$ represent the LTE reference parameters.

2.3.1.3 High Layer Split

As the split point moves further down the PS to the LLS, transmission requirements become increasingly demanding, costly, and impractical for large mobile networks. Conversely, shifting from LLS to HLS substantially decreases latency and bandwidth requirements, but supports fewer centralized processing functions. Thus, finding a balance between function centralization and network feasibility is crucial. For instance, Option 1 may not efficiently accommodate features requiring cell coordination [59]. The 3GPP RAN3 working group has determined that it is crucial for lower-layer protocols like PHY, MAC, and RLC to be co-located within the DU to ensure tight synchronization. Consequently, the RRC and PDCP layers could be shifted from the DU to the CU. Based on this, the group has selected Option 2 as a preferred HLS architecture [4, 54].

In the Option 2 architecture, RRC and PDCP functionalities are handled by the CU, while RLC, MAC, and PHY functions are managed by the DU. This separation introduces a new interface called F1, designed for sending control signaling and user plane data. Option 2 offers several benefits, including improved network integration, reduced latency, lower transmission requirements and enhanced network synchronization, in comparison to LLS options [4].

Additionally, a significant merit of HLS is its support for DC. This feature enables seamless interoperability between 5G FWA and existing 4G networks. When some capacity of the 5G-based FWA network remains unused by fixed users, it may be reallocated to boost capacity for mobile counterparts. Typically, mobile users connect to the mobile network, like 4G alternatives, which provide broad coverage in the area. With DC, these users can dynamically set up a secondary connection to a 5G FWA-based cell with available resources, thereby enhancing their achievable data rates.

In addition to improving load management, real-time performance optimization, and supporting software-defined networking (SDN)/NFV features, HLS promotes versatile and economical hardware solutions. However, it entails a comparatively intricate DU in which most RAN functionalities are executed. In this context, it can lead to increased costs and complexity in RE installation and maintenance. As a result, implementing HLS can result in bulky RE that may need to be installed on utility poles or streetlamp poles [48]. Consequently, the DL and UL bandwidths for Option 2 can be defined, respectively, as [4, 36, 56, 57]

$$R_{DL}^{PDCP-RLC} = \left(R_p^{DL} BW_s^{SY} L_s^{\text{MIMO}_{DL}} M_s^{DL}\right) + \left(8 N_{\max}^{UE} P_{rep}^{UE} C_{av}^{DL}\right), \quad (2.6a)$$

$$R_{UL}^{PDCP-RLC} = \left(R_p^{UL} BW_s^{SY} L_s^{\text{MIMO}_{UL}} M_s^{UL}\right) + \left(8 N_{\max}^{UE} P_{rep}^{UE} C_{av}^{UL}\right), \quad (2.6b)$$

where 8 is the conversion factor from bytes to bits, P_{rep}^{UE} is the percentage of UE that submit (UL or DL) requests, $C_{av}^{(\cdot)}$ represents the average content size (UL or DL), and $N_{\max}^{UE}$ denotes the maximum number of UE.

According to the fundamental 5G premises outlined by 3GPP TSG RAN WG3 [56–58], and utilizing Eqs. 2.1, 2.2, and 2.6, we assess and model the bandwidth needs for UL

Fig. 2.3 Demanded MFH capacity for various split options

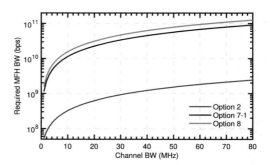

transmission, with an emphasis on comparing options 2, 7-1, and 8. Figure 2.3 illustrates the MFH bandwidth required for each option relative to the system bandwidth. For a system bandwidth of 40 MHz, Option 2 requires an MFH bandwidth of 1.224 Gbps. At an 80 MHz RF bandwidth, this requirement doubles for Option 2, demonstrating that the necessary bandwidth varies with the radio configuration. Additionally, it is evident that the bandwidth needs are significantly influenced by the precise split option. For example, at an 80 MHz system bandwidth, the demanded MFH bandwidths are 90.92 Gbps for Option 7-1 and 125.8 Gbps for Option 8. This indicates that as the split point shifts further down the PS to the LLS, the demanded MFH transport bandwidth rises and becomes increasingly demanding. For example, compared to Option 2, Option 8 needs an extra 123.378 Gbps at 80 MHz [4].

2.3.2 ITU-T RAN Architecture

The ITU-T has employed a marginally distinct architecture for 5G transport networks, which includes three logical components: the CU, DU, and RU, as illustrated in Fig. 2.4a. In this architecture, the lower and mid layer functions are split among the DU and RU. The RU handles the RF functions and, based on the specific functional split with the DU, may also manage high-PHY and low-PHY functions. As depicted in Fig. 2.4b–d, the DU, CU, and RU can be combined in various configurations to create the physical network elements. This approach offers the flexibility to support innovative applications, diverse network architectures, and transport network needs. The transport network linking the 5GC to the CU is termed backhaul, which utilizes the 3GPP NG interface. The transport network linking the CU to the DU is known as midhaul, employing the 3GPP F1 interface. The transport network between the DU and the RU is called fronthaul. Together, these networks—fronthaul, midhaul, and backhaul—are collectively referred to as xHaul [12, 36, 61].

2.3 RAN Architecture

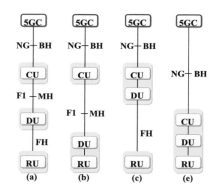

Fig. 2.4 Potential RU, DU, and CU configurations. FH: Fronthaul, MH: Midhaul and BH: Backhaul

2.3.3 Virtualized RAN

Virtualization shifts RAN functions from specialized hardware to software-based components. Virtualized RAN (vRAN) enables network functions to be managed through software platforms on standard commercial off-the-shelf (COTS) hardware. This method improves adaptability and scalability while enabling the swift deployment of new services. vRAN also supports network slicing, enabling the development of multiple virtual networks over a shared physical infrastructure. Every slice can be tailored to particular applications or user requirements, thereby optimizing overall network efficiency and service quality. In this framework, vRAN facilitates continuous and rapid evolution of the network, enabling it to adapt to the changing requirements of both novel and current services with minimal effect on CAPEX and OPEX [26, 37, 62].

2.3.4 Open RAN

In addition to the initiatives by the CPRI industry collaboration, the O-RAN Alliance revealed its commitment to advancing open RAN through RAN virtualization and interoperable interfaces. This initiative promotes open standards and interfaces to guarantee seamless interoperability between devices from various vendors. The goal is to lower costs and foster innovation by eliminating proprietary constraints. Additionally, the modular approach allows operators to integrate components from various vendors, thereby optimizing both performance and cost-efficiency [63, 64].

The O-RAN Alliance has established nine working groups that concentrated on different areas, including L2-L3 RAN protocols for HLS and L1 options (such as IEEE1914 and eCPRI) for LLS. Also, it has launched an innovative architecture for the 7.2 functional split, classified into Category A and Category B. The primary distinction between the two categories lies in the assignment of precoding functions for the DL: Category A equipment lack these functions, while Category B counterparts include them [12, 62, 63].

2.4 Conclusion

5G NR offers diverse services with varying requirements. In contrast to 4G, 5G introduces a novel network architecture designed to considerably enhance general performance. It promises multi-gigabit data rates, ultra-dense connections, and millisecond-level latency. With these advancements, 5G is poised to usher in a new era of boundless possibilities and comprehensive connectivity.

The radio signal processing stack in 5G NR operates as a sequential *service chain* of functions. These functions can be decomposed and separated with defined interfaces to achieve disaggregation. Functions requiring real-time processing are consolidated within the DU, while those that do not require real-time processing are managed within the CU.

Options 1-5 (HLS options) impose the most lenient demands on the transport network but fall short of the efficiency and performance benefits offered by more centralized schemes. On the other hand, Options 6-8 (LLS options) provide greater centralization and coordination throughout the PS, leading to improved resource utilization and enhanced radio performance. However, the LLS alternatives come with significantly stricter latency and data rate requirements, potentially restricting network deployment due to limitations in transport options and network topology. Also, they demand significantly more transport network resources, which increases deployment costs. The ideal split is determined by various technical and business factors, including network topology, fiber availability, user count, and service volume, among other things.

References

1. J. Maes, M. Peeters, M. Guenach, C. Storry, Maximizing digital subscriber line performance. Bell Labs Tech. J. **13**(1), 105–115 (2008)
2. Z. Papir, A. Simmonds, Competing for throughput in the local loop. IEEE Commun. Mag. **37**(5), 61–66 (1999)
3. N. Saba, J. Salo, K. Ruttik, R. Jäntti, Using existing base station sites for 5G millimeter-wave fixed wireless access: antenna height and coverage analysis, in *IEEE Wireless Communications and Networking Conference (WCNC)*, vol. 2024 (2024), pp. 1–6
4. I.A. Alimi, R.K. Patel, N.J. Muga, A.N. Pinto, A.L. Teixeira, P.P. Monteiro, Towards enhanced mobile broadband communications: a tutorial on enabling technologies, design considerations, and prospects of 5G and beyond fixed wireless access networks. Appl. Sci. **11**(21) (2021), https://www.mdpi.com/2076-3417/11/21/10427
5. I.A. Alimi, P.P. Monteiro, A.L. Teixeira, Analysis of multiuser mixed RF/FSO relay networks for performance improvements in Cloud Computing-Based Radio Access Networks (CC-RANs). Opt. Commun. **402**, pp. 653–661 (2017), https://www.sciencedirect.com/science/article/pii/S0030401817305734
6. Y. Chu, H. Ahmadi, D. Grace, D. Burns, Deep learning assisted fixed wireless access network coverage planning. IEEE Access **9**, 124 530–124 540 (2021)
7. A.F. Pakpahan, I.-S. Hwang, Flexible access network multi-tenancy using NFV/SDN in TWDM-PON. IEEE Access **11**, 42 937–42 948 (2023)

8. A. Ericsson, L. Falconetti, H. Olofsson, J. Edstam, T. Dahlberg, Closing the digital divide with mmWave extended range for FWA. Ericsson Technol. Rev. **2022**(11), 2–11 (2022)
9. 5G mmWave Coverage Extension Solutions, GSMA, White Paper (2022), https://www.gsma.com/futurenetworks/wp-content/uploads/2022/12/5gmmWave-coverage-extension-solutions-whitepaper_FINAL-v2.pdf. Accessed 21 Apr 2024
10. Bridging the digital divide: Extended-range millimeter-wave 5G Fixed Wireless Access, Ericsson and UScellular, Case study (2022), https://www.ericsson.com/4a9aa7/assets/local/cases/customer-cases/2022/uscellular-bridging-digital-divide.pdf. Accessed 20 Apr 2024
11. I.A. Alimi, R.K. Patel, A. Zaouga, N.J. Muga, Q. Xin, A.N. Pinto, P.P. Monteiro, Trends in cloud computing paradigms: fundamental issues, recent advances, and research directions toward 6G fog networks, in *Moving Broadband Mobile Communications Forward*, ed. A. Haidine (IntechOpen, Rijeka, 2021). https://doi.org/10.5772/intechopen.98315
12. Innovations in 5G Backhaul Technologies: IAB, HFC & Fiber, 5G Americas, White Paper (2020), https://www.5gamericas.org/wp-content/uploads/2020/06/Innovations-in-5G-Backhaul-Technologies-WP-PDF.pdf. Accessed 15 Jul 2024
13. G.K. Papageorgiou, M. Sellathurai, K. Ntougias, C.B. Papadias, A stochastic optimization approach to hybrid processing in massive MIMO systems. IEEE Wirel. Commun. Lett. **9**(6), 770–773 (2020)
14. L. Liang, W. Xu, X. Dong, Low-complexity hybrid precoding in massive multiuser MIMO systems. IEEE Wirel. Commun. Lett. **3**(6), 653–656 (2014)
15. X. Zhao, M. Li, Y. Liu, T.-H. Chang, Q. Shi, Communication-efficient decentralized linear precoding for massive MU-MIMO systems. IEEE Trans. Signal Process. **71**, 4045–4059 (2023)
16. L. Hanzo, M. El-Hajjar, O. Alamri, Near-capacity wireless transceivers and cooperative communications in the MIMO era: evolution of standards, waveform design, and future perspectives. Proceed. IEEE **99**(8), 1343–1385 (2011)
17. C. Mehlfuhrer, S. Caban, M. Rupp, Cellular system physical layer throughput: How far off are we from the Shannon bound? IEEE Wirel. Commun. **18**(6), 54–63 (2011)
18. Fixed Wireless Access with 5G Networks, 5G Americas, White Paper (2021), https://www.5gamericas.org/wp-content/uploads/2021/11/5G-FWA-WP.pdf). Accessed 21 Apr 2024
19. 5G Spectrum Public Policy Position, Huawei, Position Paper (2020). Accessed 23 Apr 2024
20. H.M. Kamdjou, D. Baudry, V. Havard, S. Ouchani, Resource-constrained extended reality operated with digital twin in industrial internet of things. IEEE Open J. Commun. Soc. **5**, 928–950 (2024)
21. A.K. Bairagi, M.S. Munir, M. Alsenwi, N.H. Tran, S.S. Alshamrani, M. Masud, Z. Han, C.S. Hong, Coexistence mechanism between eMBB and uRLLC in 5G wireless networks. IEEE Trans. Commun. **69**(3), 1736–1749 (2021)
22. H. Peng, L.-C. Wang, Z. Jian, Data-driven spectrum partition for multiplexing URLLC and eMBB. IEEE Trans. Cogn. Commun. Netw. **9**(2), 386–397 (2023)
23. L. Wan, Z. Guo, Y. Wu, W. Bi, J. Yuan, M. Elkashlan, L. Hanzo, 4G/5G spectrum sharing: efficient 5G deployment to serve enhanced mobile broadband and internet of things applications. IEEE Veh. Technol. Mag. **13**(4), 28–39 (2018)
24. P. Skokowski, K. Malon, M. Kryk, K. MaÅlanka, J.M. Kelner, P. Rajchowski, J. Magiera, Practical trial for low-energy effective jamming on private networks with 5G-NR and NB-IoT radio interfaces. IEEE Access **12**, 51 523–51 535 (2024)
25. What Is 5G NR? 5G New Radio Standard Explained, Celona, Position Paper (2022), https://www.celona.io/5g-lan/5g-nr. Accessed 23 Apr 2024
26. H.-S. Lee, S. Moon, D.-Y. Kim, J.-W. Lee, Packet-based fronthauling in 5G networks: network slicing-aware packetization. IEEE Commun. Stand. Mag. **7**(2), 56–63 (2023)

27. F.W. Vook, A. Ghosh, E. Diarte, M. Murphy, 5G new radio: overview and performance, in *2018 52nd Asilomar Conference on Signals, Systems, and Computers* (2018), pp. 1247–1251
28. V. Ramireddy, M. Grossmann, M. Landmann, G.D. Galdo, Enhancements on Type-II 5G new radio codebooks for UE mobility scenarios. IEEE Commun. Stand. Mag. **6**(1), 35–40 (2022)
29. M. Girnyk, H. Jidhage, S. Faxér, Broad beamforming technology in 5G massive MIMO. Ericsson Technol. Rev. **2023**(10), 2–6 (2023)
30. P. Bacon, Y.-T. Lee, 5G NR challenges and trends in RFFE design. Microw. J. 1–12 (2023)
31. T. Hatt, P. Jarich, E. Kolta, J. Joiner, 5G fixed wireless: a renewed playbook, GSMA Intelligence, White Paper (2021), https://data.gsmaintelligence.com/api-web/v2/research-file-download?id=60620869&file=020321-5G-fixed-wireless.pdf. Accessed 21 May 2024
32. M. Kim, S.-E. Hong, J.-H. Na, Beam selection for cell-free millimeter-wave massive MIMO systems: a matching-theoretic approach. IEEE Wirel. Commun. Lett. **12**(8), 1459–1463 (2023)
33. Z. Gu, H. Lu, P. Hong, Y. Zhang, Reliability enhancement for VR delivery in mobile-edge empowered dual-connectivity Sub-6 GHz and mmWave HetNets. IEEE Trans. Wirel. Commun. **21**(4), 2210–2226 (2022)
34. M. Abdelghaffar, T.V.P. Santhappan, Y. Tokgoz, K. Mukkavilli, A. Tingfang Ji, Subband full-duplex large-scale deployed network designs and tradeoffs. Proceed. IEEE **112**(5), 487–510 (2024)
35. K. Aldubaikhy, W. Wu, N. Zhang, N. Cheng, X. Shen, mmWave IEEE 802.11ay for 5G fixed wireless access. IEEE Wirel. Commun. **27**(2), 88–95 (2020)
36. I.A. Alimi, P.P. Monteiro, Functional split perspectives: a disruptive approach to RAN performance improvement. Wirel. Pers. Commun. **106**(1), 205–218 (2019). https://doi.org/10.1007/s11277-019-06272-7
37. Virtualized Radio Access Network-Architecture, Key technologies and Benefits, Samsung Electronics Co., Ltd., Technical Report Ver.2-1 (2019), https://images.samsung.com/is/content/samsung/p5/global/business/networks/insights/white-paper/virtualized-radio-access-network/white-paper_virtualized-radio-access-network.pdf. Accessed 22 Jul 2024
38. J. Kim, S. Bahk, Blockage-aware flow control in E-UTRA-NR dual connectivity for QoS enhancement. IEEE Access **10**, 68 834–68 845 (2022)
39. C. Pupiales, D. Laselva, I. Demirkol, Fast data recovery for improved mobility support in multiradio dual connectivity. IEEE Access **10**, 93 674–93 691 (2022)
40. 3GPP TR-38.801 R14, Study on New Radio Access Technology; Radio Access Architecture and Interfaces (2017), https://portal.3gpp.org/desktopmodules/Specifications/SpecificationDetails.aspx?specificationId=3056
41. C.-Y. Wu, H. Li, O. Caytan, J. Van Kerrebrouck, L. Breyne, J. Bauwelinck, P. Demeester, G. Torfs, Distributed multi-user MIMO transmission using real-time sigma-delta-over-fiber for next generation fronthaul interface. J. Lightwave Technol. **38**(4), 705–713 (2020)
42. M. Wong, A. Prasad, A.C.K. Soong, The security aspect of 5G fronthaul. IEEE Wirel. Commun. **29**(2), 116–122 (2022)
43. J. Zou, S. Adrian Sasu, M. Lawin, A. Dochhan, J.-P. Elbers, M. Eiselt, Advanced optical access technologies for next-generation (5G) mobile networks. J. Opt. Commun. Netw. **12**(10), D86–D98 (2020)
44. Virtualized RAN, Samsung Electronics Co., Ltd., Technical White Paper, vol. 2 (2021), https://images.samsung.com/is/content/samsung/p5/global/business/networks/insights/white-papers/0406_virtualized-ran-vol-2/Virtualized_RAN-Vol.2.pdf. Accessed 22 Jul 2024
45. I.A. Alimi, A.L. Teixeira, P.P. Monteiro, Toward an efficient C-RAN optical fronthaul for the future networks: a tutorial on technologies, requirements, challenges, and solutions. IEEE Commun. Surv. Tutor. **20**(1), 708–769 (2018)

46. J. Wang, Z. Jia, L.A. Campos, C. Knittle, Delta-sigma modulation for next generation fronthaul interface. J. Lightwave Technol. **37**(12), 2838–2850 (2019)
47. S.T. Le, T. Drenski, A. Hills, M. King, K. Kim, Y. Matsui, T. Sizer, 100Gbps DMT ASIC for hybrid LTE-5G mobile fronthaul networks. J. Lightwave Technol. **39**(3), 801–812 (2021)
48. B. Skubic, M. Fiorani, S. Tombaz, A. Furuskär, J. Mårtensson, P. Monti, Optical transport solutions for 5G fixed wireless access [Invited]. IEEE/OSA J. Opt. Commun. Netw. **9**(9), D10–D18 (2017)
49. C. Ranaweera, E. Wong, A. Nirmalathas, C. Jayasundara, C. Lim, 5G C-RAN with optical fronthaul: an analysis from a deployment perspective. J. Lightwave Technol. **36**(11), 2059–2068 (2018)
50. J. Bartelt, P. Rost, D. Wubben, J. Lessmann, B. Melis, G. Fettweis, Fronthaul and backhaul requirements of flexibly centralized radio access networks. IEEE Wirel. Commun. **22**(5), 105–111 (2015)
51. A. Pizzinat, P. Chanclou, F. Saliou, T. Diallo, Things you should know about fronthaul. J. Lightwave Technol. **33**(5), 1077–1083 (2015)
52. A.S. Thyagaturu, Z. Alharbi, M. Reisslein, R-FFT: function split at IFFT/FFT in unified LTE CRAN and cable access network. IEEE Trans. Broadcast. **64**(3), 648–665 (2018)
53. I. Ajewale Alimi, N. Jesus Muga, A.M. Abdalla, C. Pinho, J. Rodriguez, P. Pereira Monteiro, A. Luís Teixeira, *Towards a Converged Optical-Wireless Fronthaul/Backhaul Solution for 5G Networks and Beyond* (John Wiley & Sons, Ltd, 2019), ch. 1, pp. 1–29, https://onlinelibrary.wiley.com/doi/abs/10.1002/9781119491590.ch1
54. NGMN Overview on 5G RAN Functional Decomposition, NGMN Alliance, Final Deliverable (approved), Version: 1.0 (2018), https://www.ngmn.org/fileadmin/ngmn/content/downloads/Technical/2018/180226_NGMN_RANFSX_D1_V20_Final.pdf
55. SCF-159 R7, Small Cell Virtualization Functional Splits and Use Cases (2016), http://scf.io/en/documents/159_-_Small_Cell_Virtualization_Functional_Splits_and_Use_Cases.php
56. 3GPP R3-161813, Transport requirement for CU&DU functional splits options (2016), https://portal.3gpp.org/ngppapp/CreateTdoc.aspx?mode=view&contributionId=723384
57. 3GPP-WG3 R3-162102, CU-DU split: Refinement for Annex A (Transport network and RAN internal functional split) (2016), http://www.3gpp.org/DynaReport/TDocExMtg--R3-93b--31676.htm
58. 3GPP TR38.816 V15.0.0, Study on CU-DU lower layer split for NR (2017), https://portal.3gpp.org/desktopmodules/Specifications/SpecificationDetails.aspx?specificationId=3364
59. Common Public Radio Interface: Requirements for the eCPRI Transport Network, eCPRI, eCPRI Transport Network V1.0, Requirements Specification (2017), http://www.cpri.info/downloads/Requirements_for_the_eCPRI_Transport_Network_V1_0_2017_10_24.pdf
60. K. Miyamoto, S. Kuwano, T. Shimizu, J. Terada, A. Otaka, Performance evaluation of ethernet-based mobile fronthaul and wireless COMP in split-PHY processing. IEEE/OSA J. Opt. Commun. Netw. **9**(1), A46–A54 (2017)
61. M. Klinkowski, M. Jaworski, Dedicated path protection with wavelength aggregation in 5G packet-optical Xhaul access networks. J. Lightwave Technol. **41**(6), 1591–1602 (2023)
62. M. Klinkowski, Optimized planning of DU/CU placement and flow routing in 5G packet Xhaul networks. IEEE Trans. Netw. Serv. Manag. **21**(1), 232–248 (2024)
63. P. Baguer, G.M. Yilma, E. Municio, G. Garcia-Aviles, A. Garcia-Saavedra, M. Liebsch, X. Costa-Pérez, Attacking O-RAN interfaces: threat modeling, analysis and practical experimentation. IEEE Open J. Commun. Soc. **5**, 4559–4577 (2024)
64. E. Municio, G. Garcia-Aviles, A. Garcia-Saavedra, X. Costa-Pérez, O-RAN: analysis of latency-critical interfaces and overview of time sensitive networking solutions. IEEE Commun. Stand. Mag. **7**(3), 82–89 (2023)

Transport Network Architectures and Requirements

3.1 Introduction

5G is set to transform connectivity with its potential for ultra-high speeds, low latency, and extensive device connectivity. Unlike earlier generations, 5G is engineered to accommodate a diverse range of applications. The effective rollout of 5G relies on a robust and advanced transport network capable of meeting its stringent demands. Achieving these goals requires substantial improvements in the underlying transport infrastructure that interlinks various network components, such as BSs, data centers, and edge computing nodes [1–4].

In reality, the real-world transport requirements for 5G, along with their associated components and functionalities, will vary based on multiple factors, such as channel bandwidth per band, the number of spectrum bands, the highest supported modulation scheme, the number of MIMO layers, the specific use cases the network must support, and the number of deployed transmit and receive antennas [5].

3.2 Transport Network Architecture

Today, the Internet profoundly impacts daily life, transforming once-visionary concepts like self-driving cars and automated industrial robots into reality with the emergence of 5G and 6G technologies [4, 6, 7]. As digital applications demand robust network infrastructure, the transition from copper to fiber and from bits per second (bps) to gigabits or terabits per second (Gbps or Tbps) has marked significant evolution for telecommunication operators. This evolution is set to continue. Access networks, the final segment connecting end users, use both wireline and wireless technologies to provide high-density connections. The challenge lies in efficiently distributing digital signals to each customer while maintaining cost-effectiveness. The study of access networks is both complex and intriguing, as it involves devising various efficient solutions [8–10].

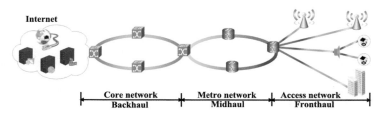

Fig. 3.1 Hierarchical inter networking model of a communications network

Access networks are part of the hierarchical inter-networking model [11], with transport network architecture typically comprising three primary segments: core, metro, and access networks, as illustrated in Fig. 3.1.

3.2.1 Core Network

The core network, also referred to as the backbone network, handles a broad spectrum of services, such as Internet content delivery, telephone traffic, and packet routing between sub-networks, with redundant connections and high-speed across various metro networks. As illustrated in Fig. 3.1, core networks typically span distances of 1,500–2,500 km, connecting major cities and sometimes extending across national borders. Modern industrial core servers offer switching capacities of up to 80 Tbps per chassis, equipped with 100 and 400 Gbps ports. To ensure high redundancy, core networks utilize a mesh topology, where each network device is interconnected with others, significantly reducing the risk of network outages [11].

3.2.2 Metro Network

The metro network is responsible for routing traffic between core networks and access networks, as well as aggregating access network traffic for delivery to the core network. Due to its role, it is also known as the aggregation or distribution network. Typically, metro network connections span distances of 50–1,000 km. To provide redundancy while managing costs effectively, a ring topology is commonly employed, offering a more economical alternative to mesh solutions [11].

3.2.3 Access Network

The access network, often known as the *last mile* or *last kilometer*, connects all mobile and fixed end-users. It is situated closest to the end-users and aims to serve as many subscribers as possible. This network includes a range of technologies such as copper wires (DSL), coaxial cables, fiber optics (FTTH or FTTB), and wireless connections (Wi-Fi or cellular networks), each designed to meet specific service needs and deployment conditions [12–14]. As illustrated in Fig. 3.1, access networks cater to diverse customer types, including individuals, enterprises, and both private and public entities. Depending on client density and service requirements, various fixed topologies are employed. Given the complexity of access networks, operators must carefully select adaptive technologies to meet bitrate, latency, jitter, and other requirements, as well as manage QoS configurations effectively. Additionally, for extensive fiber implementations in the final 10–20 km, operators must balance performance and cost considerations [1, 11].

3.3 Transport Network Architecture Evolution

As we transition from 4G/LTE to the 5G NR transport architecture, a key modification is the division of the original Baseband Unit (BBU) function into three components: CU, DU, and Remote Radio Head (RRH), also referred to as Remote Radio Unit (RRU). This reform is driven by several factors. For instance, it supports enhanced RAN virtualization and reduces the required fronthaul line rates while still satisfying latency requirements [15–20].

In 4G networks, mobile backhaul encompasses the segment of the telecommunications network that carries traffic from BSs at cell towers to the closest traffic switching center. While various backhaul solutions are accessible, including optical fiber, optical wireless communications (OWC), microwave, and copper. The emergence of pre-5G and 5G technologies is leading mobile operators to increasingly favour optical fiber as their preferred medium. The 5G era introduces new radio access architectures like C-RAN, and utilizes protocols like CPRI/eCPRI to link numerous RUs at street level to a centralized cloud-based baseband unit (DU/CU) at the macro BS. Although optical fiber continues to be the preferred option for fronthaul network, 5G introduces heightened demands for latency, jitter, scalability, and bandwidth [16, 17, 21]. These challenges are addressed by the xHaul architecture, which combines backhaul, midhaul, and fronthaul into a unified transport network, thereby reducing both CAPEX and OPEX [1, 22–25]. This xHaul architecture, recommended in [15], is illustrated in Fig. 3.2 and offers a practical solution for meeting HBB requirements and bridging the digital divide.

3.4 FWA Transport Network

The selection of the most suitable transport scheme relies on factors like the current copper or fiber infrastructure and the site's configuration [26]. FWA delivers connectivity through wireless links in the last mile. For the connections linking the core network to the BSs (i.e., the transport network), a variety of transport solutions are available, encompassing both wireline and wireless options. There is no one-size-fits-all solution for the transport

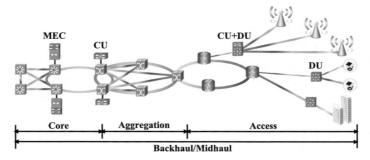

Fig. 3.2 Unified 5G transport network architecture (Adapted from [15])

network; operators ought to consider the accessible alternatives along with their respective benefits and drawbacks. Table 3.1 offers a high-level comparison of the different options [26–28].

3.4.1 Fronthaul Network Interfaces

Various fronthaul network interfaces have been established, with the CPRI and eCPRI being the most prevalent [30–32]. These interfaces are defined by the CPRI industry cooperation. The CPRI specification, established in 2003, serves as a standard interface between the BBU and the RRH. In this context, the RRH corresponds to an RU featuring an Option 8 functional split (i.e., RF/Low PHY), while the BBU represents a unified DU and CU. CPRI was developed to transmit digitized time-domain samples of the baseband signal between the BBU and the RRH. This setup offers benefits such as simplified RRH equipment, simpler operation, and reduced energy consumption, lower maintenance costs at the network edge. eCPRI was developed to address the limitations of CPRI in 5G systems, where extensive channel bandwidths and huge antenna numbers make CPRI impractical. eCPRI alleviates the requirements on transport capacity through flexible functional decomposition and simplifies the intricacy of the RU [33, 34]. It provides a tenfold decrease in the demanded data rate in comparison to CPRI. Also, it supports packet-based transport technologies like Ethernet [16, 17, 35–37].

Meanwhile, the IEEE 1914 has been developing the Next Generation Fronthaul Interface (NGFI). This initiative consists of two components: IEEE 1914.1, which focuses on standards for packet-based fronthaul transport networks, and IEEE 1914.3, which addresses Radio over Ethernet (RoE) encapsulation and mappings for DU/CU splits 7.1/7.2 and 8 [5, 17, 18, 38].

MNOs upgrading to emerging NR technology at current LTE sites face the difficulty of managing both CPRI and eCPRI during the transition period. Once every traffic at the cell site is converted to eCPRI, they can shift to packet-switched networks for the transport layer. This transition can be supported by novel protocols like TSN (IEEE 802.1CM), which

3.4 FWA Transport Network

Table 3.1 Possible FWA transport solutions

System	Solution	Advantages	Disadvantages	References
Optical	P2P Fibre (Grey Optics)	• Affordable optics with support for low latency and high capacity	• Requires a deployment with abundant fiber	[26–29]
	TWDM-PON (e.g., XG-PON, GE-PON, NG-PON2)	• Low cost potential • Possibility of system reuse for both FTTH and FWA clients	• Limited support for low latency and limited capacity (<10 Gbps) • Restricts available RAN deployment options and services, particularly in terms of low latency	[26–29]
	WDM-PON (e.g., WS-WDM-PON, WR-WDM-PON)	• Specialized scheme for RAN transport and customizable for the specific RAN deployment needs	• Restricted ability to reuse existing FTTH infrastructure • Possible challenges for migrating customers to FTTH in the future	[26–29]
	P2P WDM Overlay (e.g., NG-PON2)	• Utilization of possibly existing fiber infrastructure to offer P2P links for mobile transport • Capability to handle low latency and high capacity services	• High footprint and costs	[26–29]
Active	Ethernet (e.g., CPRI over Ethernet), OTN	• Low-cost optics and support for low latency and high capacity	• Needs a deployment with abundant fiber	[26–28]
Wireless	In-band wireless (e.g., LTE, 5G)	• Low-cost implementation	• Spectrum is allocated between transport and access	[26–28]
	Out-of-band wireless (e.g., FSO, microwave)	• Lower deployment costs compared to fiber, though requiring more effort than in-band wireless	• Dependent on the specific solution or spectrum	[26–28]

are designed to provide precise synchronization and low latency for fronthaul traffic [36, 38–40].

Table 3.2 outlines the latency requirements, data rate, and distance limitations for the F1, NG, CPRI, and eCPRI interfaces, as depicted in Fig. 3.3. The upper section of the illustration displays the functions being performed at the radio antenna site, while the lower section shows the functions carried out at the central site. The transport needs are determined by a radio site featuring 3 sectors with 64 transmit/receive chains utilizing MU-MIMO, 100 MHz channel bandwidth, 256 QAM, and 16 MIMO layers. This figure, along with Table 3.2,

Table 3.2 Specifications and distance constraints for different interfaces

	Interfaces			
	NG	F1	eCPRI	CPRI
Data rate	10 Gbps	10 Gbps	100 Gbps	1 Tbps
Latency	<10 ms	<5 ms	<200 µs	<100 µs
Distance	200–500 km	100–400 km	<20 km	<10 km

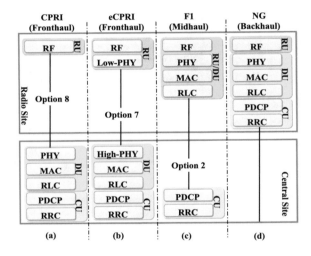

Fig. 3.3 5G NR functional splits for **a** CPRI, **b** eCPRI, **c** F1, and **d** NG interfaces

highlight the substantial differences in latencies and data rates required for the different fronthaul, midhaul, and backhaul interfaces [5, 41].

3.4.2 RAN Split Options for FWA

FWA introduces new challenges for cell site connections. Unlike traditional macro implementations, FWA may necessitate up to ten times more cells and cell site connections, significantly increasing the strain on the backhaul network. Additionally, the required inter-site distance (ISD) can range from some hundred meters to a few kilometers, based on the specifics of the 5G radio deployment [26].

By means of 5G, various functional splits are introduced to support novel deployment scenarios for RAN functions at different sites. Since FWA utilizing 3GPP RATs fundamentally adheres to the 3GPP RAN architecture, the radio functional split options designed for mobile networks are as well applicable [26, 27]. It is important to note that the precise demands on the transport network are determined by the RAN deployment strategy and the interfaces transported over the network [26].

3.5 Conclusion

Table 3.3 Interface capacities of different RAN splits

RAN split	Protocol stack implementation per site	Parameters that determine capacity	Required transport capacity per site (Gbps)	References
CPRI (4G)	Only RF is implemented	Number of antennas, bandwidth, SNR, overhead, traffic load	40–100	[26, 27]
Evolved CPRI	Only RF is implemented, with plans to reduce the transport capacity through evolution	Number of antennas, bandwidth, SNR, overhead, traffic load	40	[26, 27]
Split-PHY	Lower PHY and RF are implemented	Number of antennas, bandwidth, SNR, overhead, traffic load	10–25	[26, 27]
MAC-PHY split	Lower PHY, upper PHY, and RF implemented	Number of antennas, bandwidth, SNR, traffic load	5–10	[26, 27]

Generally, deploying FWA necessitates enhanced array antennas to enable beamforming and MU-MIMO, which are essential for achieving the demanded capacities and peak rates for household access[41–43]. This requirement, consequently, impacts the interface capacities of various RAN splits, as detailed in Table 3.3. Pertaining to lower splits like CPRI, the antenna setup in FWA has the potential to demand exceedingly high transport bitrates, that is impractical for the access segment. Alternatively, higher splits (such as MAC-PHY) are more feasible, as they allow transport bitrates to remain below 10G per site [26]. Thus, a trade-off exists between the complexity of site deployment and transport capacity per site [27].

3.5 Conclusion

The split in 5G RAN architecture is crucial for developing FWA solutions because it allows for dynamic placement of functions, comprising those within the RAN, throughout the access network to meet a number of requirements. Functional nodes can be implemented on hardware pools that include both specialized and general-purpose processors, providing the flexibility needed to adapt to future demands for capacity, latency, and evolving applications like AR and VR in residential settings.

Furthermore, the FWA transport network can utilize a variety of transport solutions, both wireline and wireless. While there is no one-size-fits-all option for the transport network, operators should carefully consider the available options and their respective benefits and drawbacks.

References

1. J. Prados-Garzon, T. Taleb, M. Bagaa, Optimization of flow allocation in asynchronous deterministic 5G transport networks by leveraging data analytics. IEEE Trans. Mob. Comput. **22**(3), 1672–1687 (2023)
2. I.A. Alimi, R.K. Patel, A. Zaouga, N.J. Muga, Q. Xin, A.N. Pinto, P.P. Monteiro, Trends in cloud computing paradigms: fundamental issues, recent advances, and research directions toward 6G fog networks, in *Moving Broadband Mobile Communications Forward*, ed. by A. Haidine (IntechOpen, Rijeka, 2021), ch. 1. https://doi.org/10.5772/intechopen.98315
3. A. Garcia-Saavedra, J.X. Salvat, X. Li, X. Costa-Perez, WizHaul: on the centralization degree of cloud RAN next generation fronthaul. IEEE Trans. Mob. Comput. **17**(10), 2452–2466 (2018)
4. R.I. Abd, D.J. Findley, K. Soon Kim, Hydra radio access network (H-RAN): multi-functional communications and sensing networks, initial access implementation, task-1 approach. IEEE Access **12**, 76 532–76 554 (2024)
5. Innovations in 5G Backhaul Technologies: IAB, HFC & Fiber, 5G Americas, White Paper (2020), https://www.5gamericas.org/wp-content/uploads/2020/06/Innovations-in-5G-Backhaul-Technologies-WP-PDF.pdf. Accessed 15 Jul 2024
6. T. Harada, Y. Sakai, T. Shimada, T. Yoshida, In-service X-Haul reconfiguration for on-demand mobile edge computing (MEC) service. IEICE Commun. Express **12**(12), 637–639 (2023)
7. F. Kooshki, M.A. Rahman, M.M. Mowla, A.G. Armada, A. Flizikowski, Efficient radio resource management for future 6G mobile networks: a cell-less approach. IEEE Netw. Lett. **5**(2), 95–99 (2023)
8. F. Rambach, B. Konrad, L. Dembeck, U. Gebhard, M. Gunkel, M. Quagliotti, L. Serra, V. Lopez, A multilayer cost model for metro/core networks. J. Opt. Commun. Netw. **5**(3), 210–225 (2013)
9. C. Bhar, A. Mitra, G. Das, D. Datta, Enhancing end-user bandwidth using content sharing over optical access networks. J. Opt. Commun. Netw. **9**(9), 756–772 (2017)
10. T. Tachibana, K. Sawada, H. Fujii, R. Maruyama, T. Yamada, M. Fujii, T. Fukuda, Open multi-access network platform with dynamic task offloading and intelligent resource monitoring. IEEE Commun. Mag. **60**(8), 52–58 (2022)
11. M. Wang, Virtualization (SDN/NFV) in optical access networks: contributions for enhanced fixed-mobile convergence, PhD thesis, Ecole nationale supérieure Mines-Télécom Atlantique, NNT : 2022IMTA0319ff. fftel-03881950 (2022), https://theses.hal.science/tel-03881950v1/file/2022IMTA0319_Wang-Minqi.pdf
12. A.F. Pakpahan, I.-S. Hwang, Flexible access network multi-tenancy using NFV/SDN in TWDM-PON. IEEE Access **11**, 42 937–42 948 (2023)
13. F. Mazzenga, R. Giuliano, F. Vatalaro, Sharing of copper pairs for improving DSL performance in FTTx access networks. IEEE Access **7**, 6637–6649 (2019)

References

14. M. Nakamura, H. Ueda, S. Makino, T. Yokotani, K. Oshima, Proposal of networking by PON technologies for full and ethernet services in FTTx. J. Lightwave Technol. **22**(11), 2631–2640 (2004)
15. Transport network support of IMT-2020/5G, ITU-T, Technical Report GSTR-TN5G (2018-02) (2018), https://www.itu.int/dms_pub/itu-t/opb/tut/T-TUT-HOME-2018-PDF-E.pdf. Accessed 15 Jul 2024
16. J. Zou, S. Adrian Sasu, M. Lawin, A. Dochhan, J.-P. Elbers, M. Eiselt, Advanced optical access technologies for next-generation (5G) mobile networks. J. Opt. Commun. Netw. **12**(10), D86–D98 (2020)
17. I. Chih-Lin, H. Li, J. Korhonen, J. Huang, L. Han, RAN revolution with NGFI (xhaul) for 5G. J. Lightwave Technol. **36**(2), 541–550 (2018)
18. I. Chih-Lin. Y. Yuan, J. Huang, S. Ma, C. Cui, R. Duan, Rethink fronthaul for soft RAN. IEEE Commun. Mag. **53**(9), 82–88 (2015)
19. Y. Zhu, W. Hu, Optical access networks for fixed and mobile applications [Invited Tutorial]. J. Opt. Commun. Netw. **16**(2), A118–A135 (2024)
20. M. Wong, A. Prasad, A.C.K. Soong, The security aspect of 5G fronthaul. IEEE Wirel. Commun. **29**(2), 116–122 (2022)
21. J. Wang, Z. Jia, L.A. Campos, C. Knittle, Delta-sigma modulation for next generation fronthaul interface. J. Lightwave Technol. **37**(12), 2838–2850 (2019)
22. X-Haul Architecture for 5G Transport, Tejas Networks, White Paper (2020), https://www.tejasnetworks.com/wp-content/uploads/2024/02/Tejas-whitepaper-5G-xhaul-architecture-final.pdf. Accessed 15 Jul 2024
23. P. Öhlén, B. Skubic, A. Rostami, M. Fiorani, P. Monti, Z. Ghebretensaé, J. Mårtensson, K. Wang, L. Wosinska, Data plane and control architectures for 5G transport networks. J. Lightwave Technol. **34**(6), 1501–1508 (2016)
24. I.A. Alimi, P.P. Monteiro, Functional split perspectives: a disruptive approach to RAN performance improvement. Wirel. Pers. Commun. **106**(1), 205–218 (2019). https://doi.org/10.1007/s11277-019-06272-7
25. I. Ajewale Alimi, N. Jesus Muga, A.M. Abdalla, C. Pinho, J. Rodriguez, P. Pereira Monteiro, A. Luís Teixeira, *Towards a Converged Optical-Wireless Fronthaul/Backhaul Solution for 5G Networks and Beyond* (John Wiley & Sons, Ltd, 2019), ch. 1, pp. 1–29, https://onlinelibrary.wiley.com/doi/abs/10.1002/9781119491590.ch1
26. K. Laraqui, S. Tombaz, A. Furuskär, B. Skubic, A. Nazari, E. Trojer, Fixed wireless access on a massive scale with 5G. Ericsson, Technol. Rev. **94** (2017), https://www.ericsson.com/assets/local/publications/ericsson-technology-review/docs/2017/2017-01-volume-94-etr-magazine.pdf
27. Fixed Wireless Access: Economic Potential and Best Practices, GSMA, White Paper (2018), https://www.gsma.com/solutions-and-impact/technologies/networks/wp-content/uploads/2018/08/Fixed-Wireless-Access-economic-potential-and-best-practices.pdf. Accessed 05 Jul 2024
28. I.A. Alimi, R.K. Patel, N.J. Muga, A.N. Pinto, A.L. Teixeira, P.P. Monteiro, Towards enhanced mobile broadband communications: a tutorial on enabling technologies, design considerations, and prospects of 5G and beyond fixed wireless access networks. Appl. Sci. **11**(21) (2021), https://www.mdpi.com/2076-3417/11/21/10427
29. C. Ranaweera, P. Monti, B. Skubic, E. Wong, M. Furdek, L. Wosinska, C.M. Machuca, A. Nirmalathas, C. Lim, Optical transport network design for 5G fixed wireless access. J. Lightwave Technol. **37**(16), 3893–3901 (2019)
30. H. Zeng, X. Liu, S. Megeed, A. Shen, F. Effenberger, Digital signal processing for high-speed fiber-wireless convergence [invited]. J. Opt. Commun. Netw. **11**(1), A11–A19 (2019)

31. J. Maes, S. Bidkar, M. Straub, T. Pfeiffer, R. Bonk, Efficient transport of enhanced CPRI fronthaul over PON [Invited]. J. Opt. Commun. Netw. **16**(2), A136–A142 (2024)
32. S.T. Le, T. Drenski, A. Hills, M. King, K. Kim, Y. Matsui, T. Sizer, 100Gbps DMT ASIC for hybrid LTE-5G mobile fronthaul networks. J. Lightwave Technol. **39**(3), 801–812 (2021)
33. Common Public Radio Interface: Requirements for the eCPRI Transport Network, eCPRI, eCPRI Transport Network V1.0, Requirements Specification, http://www.cpri.info/downloads/Requirements_for_the_eCPRI_Transport_Network_V1_0_2017_10_24.pdf (2017)
34. Common Public Radio Interface: eCPRI Interface Specification, eCPRI, eCPRI Specification V1.1 (2018), http://www.cpri.info/downloads/eCPRI_v_1_1_2018_01_10.pdf
35. L. Li, M. Bi, H. Xin, Y. Zhang, Y. Fu, X. Miao, A.M. Mikaeil, W. Hu, Enabling flexible link capacity for eCPRI-based fronthaul with load-adaptive quantization resolution. IEEE Access **7**, 102 174–102 185 (2019)
36. G. Otero Pérez, D. Larrabeiti López, J.A. Hernández, 5G new radio fronthaul network design for eCPRI-IEEE 802.1CM and extreme latency percentiles. IEEE Access **7**, 82 218–82 230 (2019)
37. IEEE Draft Standard for Local and metropolitan area networks—Time-Sensitive Networking for Fronthaul Amendment: Enhancements to Fronthaul Profiles to Support New Fronthaul Interface, Synchronization, and Syntonization Standards, in *IEEE P802.1CMde/D2.0*, pp. 1–39 (2019)
38. IEEE/ISO/IEC International Standard-Telecommunications and information exchange between information technology systems–Requirements for local and metropolitan area networks–Part 1CM:Time-sensitive networking for fronthaul—AMENDMENT 1: Enhancements to fronthaul profiles to support new fronthaul interface, synchronization, and syntonization standards, in *ISO/IEC/IEEE 8802-1CM:2019/Amd.1:2021(E)*, pp. 1–38 (2021)
39. IEEE Draft Time-Sensitive Networking for Fronthaul, in *IEEE P802.1CM/D2.0*, pp. 1–72 (2018)
40. E. Municio, G. Garcia-Aviles, A. Garcia-Saavedra, X. Costa-Pérez, O-RAN: analysis of latency-critical interfaces and overview of time sensitive networking solutions. IEEE Commun. Stand. Mag. **7**(3), 82–89 (2023)
41. X. Zhao, M. Li, Y. Liu, T.-H. Chang, Q. Shi, Communication-efficient decentralized linear precoding for massive MU-MIMO systems. IEEE Trans. Signal Process. **71**, 4045–4059 (2023)
42. M. Girnyk, H. Jidhage, S. Faxér, Broad beamforming technology in 5G massive MIMO. Ericsson Technol. Rev. **2023**(10), 2–6 (2023)
43. C.-Y. Wu, H. Li, O. Caytan, J. Van Kerrebrouck, L. Breyne, J. Bauwelinck, P. Demeester, G. Torfs, Distributed multi-user MIMO transmission using real-time sigma-delta-over-fiber for next generation fronthaul interface. J. Lightwave Technol. **38**(4), 705–713 (2020)

5G FWA Technological Improvements

4.1 Introduction

Independence from 4G infrastructure allows service providers to design 5G networks, maximizing the potential of spectrum resources. This strategic autonomy, along with immediate NR access and reduced latency, creates a more efficient FWA ecosystem. It represents a significant move towards greater efficiency, where the lower cost per gigabyte for 5G FWA is achieved, highlighting the financial viability of 5G [1].

The initial commercial 5G networks used LTE infrastructure for radio access and the core network to speed up deployment, known as 5G Non-Standalone (NSA). 5G NSA enabled service providers to increase bandwidth for users by combining 4G and 5G carriers through DC. To fully realize the capabilities and performance of 5G, service providers are now adopting 5G Standalone (SA), which features a dedicated core and a highly efficient 5G air interface, independent of existing LTE networks. In this context, 5G NR SA unlocks the full potential of 5G. The 5G NR SA software upgrade enhances user experience, expands coverage, improves network efficiency, reduces complexity, and opens up new business opportunities [1, 2].

Performance is a key focus as 5G SA optimizes NR spectrum utilization. Instant access significantly simplifies RAN and CPE devices. Unlike 5G NSA, SA operates independently of an LTE anchor, enabling ultra-fast connections and enhancing the user experience. The SA approach, which avoids power-sharing with LTE, allows for full power utilization for NR, resulting in superior UL coverage and throughput. Combined with superior Massive MIMO performance, these enhancements greatly improve throughput, contributing to an overall increase in network capacity available for FWA [1–3].

Additionally, service providers are now implementing Network and RAN Slicing to create 5G monetization opportunities, offering differentiated performance for new experience-based connectivity services that require high throughput, reliability, and low latency. With premium FWA as an example, the next wave is emerging by incorporating mmWave technology to further enhance gigabit FWA network performance and user experiences. This advancement paves the way for differentiated services and advanced use cases within gigabit FWA [1, 3]. Thus, 5G SA revolutionizes the FWA landscape by focusing on efficiency, performance, and differentiation. By maximizing spectrum utilization, optimizing service performance, and offering unique services, 5G SA enables service providers to develop new revenue streams through FWA connectivity [1].

Furthermore, the two primary components of the E2E wireless network architecture are the RAN and Network Core [4, 5]. Improvements and optimizations in these areas can lead to improved service control and enhance the user experience for fixed wireless services.

4.2 Radio Access Network

Following a decade of research, standardization, and industry collaboration, 5G technology has emerged. Unlike FWA, which was hitherto implemented with LTE and other proprietary technologies that constrained scalability, 5G offers a cohesive and standardized solution [6, 7]. The 5G RAN enhances deployment flexibility and network scalability, essential for meeting diverse upcoming performance requirements [5, 8–10].

Additionally, 5G introduces features that enhance performance and offer an increasingly consistent user experience. These features are designed to increase throughput and improve coverage, which are crucial aspects of FWA. The following sections discuss key 5G RAN capabilities and features that advance FWA implementations [3, 4].

4.2.1 Increased Spectral Efficiency

In comparison to LTE, the 5G NR standard significantly enhances spectral efficiency by boosting the data transmission capacity per unit of spectrum (Hz), due to various advancements in signaling and CPE technology enabled by 5G. For instance, enhancements such as antenna gains, massive MIMO, beamforming techniques, and sound reference signal (SRS) beam selection, which lets CPE switch between beams for optimal signal reception, play a significant role in these improvements. Also, the advanced numerology capability of 5G NR enables multi-Gbps peak throughput, which is particularly advantageous for delivering higher speeds to users in urban or densely populated areas for FWA [4, 6].

4.2.2 New Spectrum

New and expanded spectrum is becoming available in mid-band TDD and mmWave, offering wide bandwidths capable of handling large amounts of data traffic. mmWave-only SA is a new network option that enables FWA opportunities for service providers with only mmWave

4.2 Radio Access Network

spectrum. The mmWave spectrum offers expanded bandwidth, supporting greater speeds and reduced latency. While most existing 5G FWA implementations concentrate on the 3.5–3.8 GHz bands, mmWave spectrum is increasingly used by several operators worldwide to enhance capacity and performance, complementing the coverage offered by lower frequency bands [1, 6, 11, 12].

4.2.3 Beam Management Frameworks for 5G NR

A major aspect of the 5G NR architecture is its capability to function across two distinct frequency ranges: sub-6 GHz and mmWave. As the sub-6 GHz spectrum is increasingly scarce, mmWave frequency bands, which offer broader bandwidths, are expected to become more widely used. Operating above the 24 GHz spectrum, mmWave provides short-range, high-frequency waves with higher capacity [13–15].

The characteristics of signal blockage, propagation loss, and fading effects vary between mmWave and sub-6 GHz bands, presenting new challenges for system development and impacting E2E throughput and user experience quality [16–18]. Notably, throughput is a crucial factor in the success of 5G. To overcome these constraints, the 3GPP 5G NR standards introduce new features at the PHY and MAC layer to facilitate directional communications. A key feature is beam management, which facilitates the acquisition and maintenance of beams. The standards also define new initial access protocols to guarantee effective directional transmission [13, 19–21].

4.2.3.1 Beamforming

Full-dimension beamforming (in both horizontal and vertical directions) focuses signal energy towards the user, enhancing throughput and addressing coverage challenges. Depending on the frequency band and implementation complexity, 5G RUs utilize digital, analog, or hybrid beamforming techniques. Typically, mmWave band RUs employ hybrid or analog beamforming, whereas mid/low band RUs use digital beamforming. As illustrated in Fig. 4.1, 5G hybrid beamforming, particularly in mmWave frequencies, combines digital precoding with analog beamforming to mitigate the significant path loss (PL) at high frequencies while keeping RUs complexity manageable. For low and mid-band frequency implementations, gNB RUs can incorporate a larger number of antenna elements to enable complete digital beamforming, offering improved beam control and user beam multiplexing [4, 22–24].

4.2.3.2 Beam Management

4G LTE UE occasionally assesses the radio link to gauge the channel quality of its serving eNodeB. This process allows the UE to verify whether the network can maintain a satisfactory level of link quality. If the link quality falls below a specified threshold, the UE reports a radio link failure and initiates a higher-layer reconnection procedure. This procedure

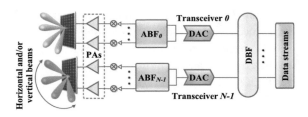

Fig. 4.1 Hybrid beamforming architecture

involves starting a new cell re-selection process, which is more time-consuming and leads to a reduction in the total data rate [19].

As 5G NR advances into the mmWave spectrum, where hybrid beamforming is commonly used at both BSs and UE sides, network management grows increasingly complex. However, highly directional beamforming architectures with a greater number of antenna elements are crucial for achieving higher data rates. Specialized procedures are necessary for UE to establish and maintain a connection, even in scenarios involving mobility [19].

mmWave technology supports directional communication with a greater number of antenna elements, providing increased beamforming gain to offset propagation loss. Nevertheless, directional links require accurate alignment of beams between the gNB and the UE. This necessitates efficient beam management, where the gNB and UE continuously determine the most effective beams to use at any specific moment [19]. To enhance coverage in higher frequency channels, especially mmWave, users are monitored and served with targeted beams in both UL and DL directions. The 5G beam management processes assist in optimizing and tracking these user beams [4, 13, 20].

4.2.4 Improved Channel State Information Mechanisms

5G introduces advanced Channel State Information Reference Signal (CSI-RS) feedback mechanisms that enhance both single-user MIMO (SU-MIMO) and MU-MIMO systems. The gNodeB transmits a DL CSI-RS signal, which the device uses to assess the channel. The device then provides feedback on the preferred codebook (quantized beam) to the gNodeB. This information allows the gNodeB to create a targeted beam for the FWA user, improving performance. A codebook is a type of matrix with complex-valued elements that transforms the data bits (PDSCH) into another set of data, which is then mapped to each antenna port. The 3GPP defines two types of CSI codebooks: Type I, generally used for SU-MIMO, and Type II, which offers more detailed and adaptable feedback suitable for both SU-MIMO and MU-MIMO [4, 25–28]. The key difference between Type I and Type II is that Type I selects a single specific beam from a set of beams, while Type II selects a set of beams and linearly combines all the beams within that cluster [29–31].

4.2.5 Multi-antenna Systems

Massive MIMO is essential in driving the advancement of 5G technology. Its beamforming, SU-MIMO, and MU-MIMO solutions significantly enhance the RAN's user experience, capacity, and coverage [11, 24, 32–34]. In this context, FWA users can achieve substantially higher upload speeds and increased capacity through UL SU-MIMO and CA. This setup necessitates devices equipped with three transmit antennas (3Tx), which is particularly advantageous for FWA service providers. The larger form factors of FWA devices simplify device output power constraints and the integration of additional radio frequency components, making them more practical compared to smartphones [1, 35, 36].

Additionally, FWA deployments with favorable LoS propagation between high tower radio sites and outdoor rooftop-mounted CPEs can achieve significantly extended mmWave cell ranges—up to several kilometers—compared to smartphones. This capability demands high-power radios, robust CPEs, and mmWave extended-range software to manage the increased propagation time over these greater distances [1].

When the system must deliver high DL traffic services, such as FWA, to multiple users simultaneously, its cell throughput and spectral efficiency can be enhanced through DL MU-MIMO (or spatial division multiple access). This technique assigns distinct groups of beams to each user. For instance, in DL SU-MIMO, beamforming weights for each layer are derived from SRS response or Pre-Coding Matrix Indicator feedback, allowing multiple layers to be allocated to a single user. Though the combined beams of DL SU-MIMO layers are not fully orthogonal, the entire signals from these layers are directed to the same user, and inter-layer interference can be reduced using innovative receiver techniques. In DL MU-MIMO, incorrectly orthogonal beams can create multi-user interference, making it challenging for the receiver of each user to cancel out this interference. Therefore, the beamforming weights must be selected carefully to optimize the beam gain toward the intended user while minimizing leakage power that affects other users receiving DL signals simultaneously on the same frequency resources [4, 35, 36].

MU-MIMO is particularly advantageous for FWA because the geographical separation of FWA UEs allows the gNB to generate superior and more orthogonal beams. This enables the simultaneous scheduling of more users on the same physical resources, thereby enhancing spectral efficiency [4]. However, using MU-MIMO exclusively can decrease the spectral efficiency for users at the cell edge. This is due to residual inter-user interference from practical multi-user beamforming and reduced transmit power per user. Consequently, it should be possible to configure the CPE-specific transmission mode to allow dynamic switching between SU-MIMO and MU-MIMO. This would help balance the spectral efficiency for cell edge users and the average spectral efficiency for all users in the cell [32, 35].

4.2.6 Downlink Coverage Utilization Expansion

Higher frequency band implementations are often limited by UL coverage, meaning that while DL coverage can extend beyond the UL link budget, the UL is more restricted. To make full use of DL capabilities beyond the UL, the FWA device can employ CA or DC with a lower frequency band. CA is crucial for optimizing 5G deployments as it maximizes the use of available spectrum and improves the user experience. It is also an effective method for extending coverage. With CA, the FWA equipment can leverage the UL of the lower frequency band while aggregating the DL from both lower and higher frequency bands, thus enhancing coverage and user-perceived throughput [1, 4].

4.2.7 Uplink Performance Improvement

Unlike mobile devices, FWA equipment can efficiently utilize multiple RF chains and antennas for both transmission and reception. With antennas that have higher gain, FWA equipment are capable of supporting multi-CA and multi-layer MIMO, particularly enhancing UL performance [4].

4.2.8 Service Differentiation

As 5G technology progresses, the variety and number of vertical services are expanding, each with its unique service level requirements. Service differentiation presents opportunities for 5G monetization through new experience-based connectivity services that demand high throughput, reliability, and low latency. 5G RAN slicing is a crucial enabler for these monetization opportunities. It involves slicing the network according to various service demands, allowing for efficient creation of user-to-application connections through logical networks supported by shared physical infrastructure. This software solution facilitates service differentiation and ensures performance for a range of new services while optimizing the use of the common 5G RAN. It supports dynamic radio resource allocation and prioritization across different slices, ensuring the fulfillment of SLA [1, 4, 37, 38].

Additionally, RAN-level network slicing features such as admission control, slice-based scheduling, and prioritization, as well as E2E slice-specific data management, enable the fulfillment of new application service requirements and open up new revenue opportunities for mobile network operators [4, 39].

Many RAN features are designed to be slice-aware, enabling the differentiation of services, functions, and behaviors in terms of configuration and observability. For instance, these features can optimize parameter settings to enhance FWA performance, direct FWA subscribers to specific frequency bands, and offer visibility into the FWA UE group. FWA typically involves high-performance devices in fixed locations with generally favorable RF

4.3 5G Core Network

conditions, allowing for targeted enhancements to optimize performance in these specific scenarios [1, 39].

4.2.9 Higher Order Modulation

The 5G standard accommodates higher-order modulation schemes (1024QAM for DL and 256QAM for UL) to achieve improve overall network coverage and greater peak throughputs. However, the benefits of these advanced modulation schemes are primarily realized close to the gNB. For FWA devices equipped with additional antenna gain, advanced modulation becomes increasingly feasible and is capable of delivering enhanced throughput to end users [4, 40, 41].

4.2.10 Dynamic TDD and Interference Mitigation

In TDD macro network implementations, a consistent TDD UL/DL ratio is typically utilized throughout the region to prevent interference between UL and DL transmissions among sites. However, 5G standards enable real-time dynamic adjustments to the TDD UL/DL ratio, allowing separate gNBs to modify the ratio based on traffic demands. To manage interference in dynamic TDD operations, 3GPP has established precise techniques and signaling methods to identify, isolate, and mitigate such interference between gNBs. Through coordinated communication, gNBs can effectively avoid interference. This capability can enhance UL and DL symmetry for both HBB and enterprise services [42, 43].

4.3 5G Core Network

While 5G does not overhaul the entire network architecture, it introduces significant advancements compared to previous generations. The 5GC network utilizes a service-based architecture (SBA), offering remarkable flexibility and scalability. This design allows for the integration of new functionalities and services without altering the current architecture [44]. The SBA architecture facilitates the adoption of innovative technologies like SDN and NFV, which offer adaptable connectivity [5, 37, 45, 46]. These technologies enable network slicing, allowing for on-demand, dedicated QoS and network access for users. Unlike earlier generations, 5G integrates seamlessly with web applications and Internet infrastructure. It avoids proprietary standards and wholly adopts Internet protocols such as HTTP/2.0 and TCP/IP. In terms of security, 5G primarily uses Transport Layer Security (TLS) for securing transport layer communications and OAuth2.0 for managing authorizations between network functions, with specific exceptions such as Protocol for N32 Interconnect Security (PRINS) used in roaming scenarios [38, 44].

The implementation of SBA and the complete integration of Internet protocols not only promise the transformative potential of 5G but also position it as a foundation for both current and future applications [44]. Additionally, 5G is characterized by improvements in core network services, as specified by 5G NR, which will drive higher throughput and data rates. These enhancements will also lead to substantial enhancements in connection density, latency, reliability, energy efficiency, content control, and data management—key features necessary for FWA. The capabilities and features of the 5GC Network that will optimize the user experience for FWA customers are detailed in [4].

4.4 Conclusion

5G RAN will enhance deployment flexibility and network scalability to meet the performance requirements of different use cases. The advanced features of 5G will improve key aspects of FWA by offering better coverage and increased throughput for users. Moreover, the ability to use software for dynamic configuration of core and service networks is crucial for achieving the flexibility needed to deploy FWA solutions on a unified wireline and mobile infrastructure. This implies that features, functions, and operational capabilities developed for mobile networks can be utilized in FWA when appropriate. Examples include optimized content delivery through blind cache, network sharing models, and unbundling via Mobile Virtual Network Operators (MVNOs) and further strategies.

Furthermore, 5G SA enables MNOs to distinguish themselves in a competitive market by enhancing their FWA offerings. By utilizing the advanced features of 5G SA, providers can implement sophisticated monetization strategies, including network slicing and speed differentiation. This approach maximizes FWA revenue and sets the stage for offering customized services with specific performance guarantees. Whether catering to residential users, businesses, or online gamers, service providers can tailor FWA services for optimal performance, thereby improving the overall customer experience.

References

1. 5G offers a future-proof platform for FWA growth, in *Ericsson, Fixed Wireless Access Handbook Extracted version, Insight 6 of 6* (2024), https://www.ericsson.com/4ade15/assets/local/reports-papers/further-insights/doc/fwa_insights_6_offers_extracted.pdf. Accessed 05 Jul 2024
2. G. Liu, Y. Huang, Z. Chen, L. Liu, Q. Wang, N. Li, 5G deployment: standalone vs. non-standalone from the operator perspective. IEEE Commun. Mag. **58**(11), 83–89 (2020)
3. U. Gustavsson, P. Frenger, C. Fager, T. Eriksson, H. Zirath, F. Dielacher, C. Studer, A. Pärssinen, R. Correia, J.N. Matos, D. Belo, N.B. Carvalho, Implementation challenges and opportunities in beyond-5G and 6G communication. IEEE J. Microwaves **1**(1), 86–100 (2021)
4. Fixed Wireless Access with 5G Networks, 5G Americas, White Paper (2021), https://www.5gamericas.org/wp-content/uploads/2021/11/5G-FWA-WP.pdf. Accessed 21 Apr 2024

References

5. K. Laraqui, S. Tombaz, A. Furuskär, B. Skubic, A. Nazari, E. Trojer, Fixed Wireless Access on a Massive Scale with 5G. Ericsson, Technol. Rev. **94** (2017), https://www.ericsson.com/assets/local/publications/ericsson-technology-review/docs/2017/2017-01-volume-94-etr-magazine.pdf
6. F. Agnoletto, A. Goel, P. Castells, The 5G FWA opportunity: disrupting the broadband market, GSMA Intelligence, White Paper (2021), https://data.gsmaintelligence.com/api-web/v2/research-file-download?id=66289674&file=141021-5G-FWA-Opportunity.pdf. Accessed 05 Jul 2024
7. T. Harada, Y. Sakai, T. Shimada, T. Yoshida, In-service X-Haul reconfiguration for on-demand mobile edge computing (MEC) service. IEICE Commun. Express **12**(12), 637–639 (2023)
8. R.I. Abd, D.J. Findley, K. Soon Kim, Hydra radio access network (H-RAN): multi-functional communications and sensing networks, initial access implementation, Task-1 approach. IEEE Access **12**, 76 532–76 554 (2024)
9. L. Iiyambo, G. Hancke, A.M. Abu-Mahfouz, A survey on NB-IoT random access: approaches for uplink radio access network congestion management. IEEE Access **12**, 95 487–95 506 (2024)
10. F. Kooshki, M.A. Rahman, M.M. Mowla, A.G. Armada, A. Flizikowski, Efficient radio resource management for future 6G mobile networks: a cell-less approach. IEEE Netw. Lett. **5**(2), 95–99 (2023)
11. A. Lappalainen, Y. Zhang, C. Rosenberg, Planning 5G networks for rural fixed wireless access. IEEE Trans. Netw. Serv. Manage. **20**(1), 441–455 (2023)
12. I.A. Alimi, R.K. Patel, A. Zaouga, N.J. Muga, A.N. Pinto, A.L. Teixeira, P.P. Monteiro, *6G CloudNet: Towards a Distributed, Autonomous, and Federated AI-Enabled Cloud and Edge Computing* (Springer International Publishing, Cham, 2021), pp. 251–283. https://doi.org/10.1007/978-3-030-72777-2_13
13. Y. Li, B. Gao, X. Zhang, K. Huang, Beam management in millimeter-wave communications for 5G and beyond. IEEE Access **8**, 13 282–13 293 (2020)
14. K. Vuckovic, M.B. Mashhadi, F. Hejazi, N. Rahnavard, A. Alkhateeb, PARAMOUNT: toward generalizable deep learning for mmWave beam selection using Sub-6 GHz channel measurements. IEEE Trans. Wirel. Commun. **23**(5), 5187–5202 (2024)
15. L. Yashvanth, C.R. Murthy, On the impact of an IRS on the out-of-band performance in Sub-6 GHz & mmWave frequencies. IEEE Trans. Commun. 1–1 (2024)
16. C.K. Anjinappa, S. Güvenç, Coverage hole detection for mmWave networks: an unsupervised learning approach. IEEE Commun. Lett. **25**(11), 3580–3584 (2021)
17. Y. Huang, Challenges and opportunities of Sub-6 GHz integrated sensing and communications for 5G-advanced and beyond. Chin. J. Electron. **33**(2), 323–325 (2024)
18. X. Xia, F. Wu, C. Yu, Z. Jiang, J. Xu, S.-Y. Tang, Z. Wang, Y. Yao, W. Hong, Millimeter-Wave and Sub-6-GHz aperture-shared antenna and array for mobile terminals accessing 5G/6G-enabled IoT scenarios. IEEE Internet of Things J. **11**(10), 18 808–18 823 (2024)
19. Understanding 5G Beam Management, MathWorks, White Paper (2021), https://www.mathworks.com/content/dam/mathworks/mathworks-dot-com/images/responsive/supporting/campaigns/offers/5g-beam-management-white-paper/5g-beam-management-white-paper.pdf. Accessed 03 Jul 2024
20. Q. Xue, J. Guo, B. Zhou, Y. Xu, Z. Li, S. Ma, AI/ML for beam management in 5G-advanced: a standardization perspective. IEEE Veh. Technol. Mag. 2–10 (2024)
21. I.A. Alimi, J.J. Popoola, K.F. Akingbade, M.O. Kolawole, Performance analysis of bit-error-rate and channel capacity of MIMO communication systems over multipath fading channels. Int. J. Inform. Commun. Technol. **2**, 57–63 (2013), https://ijict.iaescore.com/index.php/IJICT/article/view/1047/568

22. D. Wu, T. Shen, F. Shu, Y. Wu, L. Zhu, S. Feng, M. Huang, J. Wang, Secure hybrid analog and digital beamforming for mmWave XR communications with mixed-DAC. IEEE J. Select. Top. Signal Process. **17**(5), 995–1006 (2023)
23. S. Kamiwatari, I. Kanno, T. Hayashi, Y. Amano, RF chain-wise clustering schemes for millimeter wave cell-free massive MIMO with centralized hybrid beamforming. IEEE Access, **12**, 19 682–19 693 (2024)
24. S. Gong, C. Xing, P. Yue, L. Zhao, T.Q.S. Quek, Hybrid analog and digital beamforming for RIS-assisted mmWave communications. IEEE Trans. Wirel. Commun. **22**(3), 1537–1554 (2023)
25. F. Zhang, S. Sun, Q. Gao, W. Tang, Enhanced CSI acquisition for FDD multi-user massive MIMO systems. IEEE Access **6**, 23 034–23 042 (2018)
26. H. Lee, H. Choi, H. Kim, S. Kim, C. Jang, Y. Choi, J. Choi, Downlink channel reconstruction for spatial multiplexing in massive MIMO systems. IEEE Trans. Wirel. Commun. **20**(9), 6154–6166 (2021)
27. T. Wu, X. Yin, L. Zhang, J. Ning, Measurement-based channel characterization for 5G downlink based on passive sounding in Sub-6 GHz 5G commercial networks. IEEE Trans. Wirel. Commun. **20**(5), 3225–3239 (2021)
28. S. Mukherjee, M.S. Khan, A. Kumar Reddy Chavva, Optimizing near-Field XL-MIMO communications: advanced feedback framework for CSI. IEEE Access **12**, 89 205–89 221 (2024)
29. V. Ramireddy, M. Grossmann, M. Landmann, G.D. Galdo, Enhancements on type-II 5G new radio codebooks for UE mobility scenarios. IEEE Commun. Stand. Mag. **6**(1), 35–40 (2022)
30. R.M. Dreifuerst, R.W. Heath, Machine learning codebook design for initial access and CSI Type-II feedback in Sub-6-GHz 5G NR. IEEE Trans. Wirel. Commun. **23**(6), 6411–6424 (2024)
31. M. Chen, J. Guo, C.-K. Wen, S. Jin, G.Y. Li, A. Yang, Deep learning-based implicit CSI feedback in massive MIMO. IEEE Trans. Commun. **70**(2), 935–950 (2022)
32. C.-Y. Wu, H. Li, O. Caytan, J. Van Kerrebrouck, L. Breyne, J. Bauwelinck, P. Demeester, G. Torfs, Distributed multi-user MIMO transmission using real-time sigma-delta-over-fiber for next generation fronthaul interface. J. Lightwave Technol. **38**(4), 705–713 (2020)
33. S. Adda, T. Aureli, S. D'Elia, D. Franci, N. Pasquino, S. Pavoncello, R. Suman, Enhanced methodology to characterize 3-D power monitoring and control features for 5G NR systems embedding multi-user MIMO antennas. IEEE Trans. Instrum. Meas. **72**, 1–9 (2023)
34. K.R. Jha, N. Rana, S.K. Sharma, Design of compact antenna array for MIMO implementation using characteristic mode analysis for 5G NR and Wi-Fi 6 applications. IEEE Open J. Antenn. Propag. **4**, 262–277 (2023)
35. L. Liu, R. Chen, S. Geirhofer, K. Sayana, Z. Shi, Y. Zhou, Downlink MIMO in LTE-advanced: SU-MIMO vs. MU-MIMO. IEEE Commun. Mag. **50**(2), 140–147 (2012)
36. H. Choi, A.L. Swindlehurst, J. Choi, WMMSE-based rate maximization for RIS-assisted MU-MIMO systems. IEEE Trans. Commun. **72**(8), 5194–5208 (2024)
37. V.P. Singh, M.P. Singh, S. Hegde, M. Gupta, Security in 5G network slices: concerns and opportunities. IEEE Access **12**, 52 727–52 743 (2024)
38. J. Mongay Batalla, L.J. de la Cruz Llopis, G.P. Gómez, E. Andrukiewicz, P. Krawiec, C.X. Mavromoustakis, H.H. Song, Multi-layer security assurance of the 5G automotive system based on multi-criteria decision making. IEEE Trans. Intell. Transp. Syst. **25**(5), 3496–3512 (2024)
39. H.-S. Lee, S. Moon, D.-Y. Kim, J.-W. Lee, Packet-based fronthauling in 5G networks: network slicing-aware packetization. IEEE Commun. Stand. Mag. **7**(2), 56–63 (2023)
40. J. Wang, Z. Yu, K. Ying, J. Zhang, F. Lu, M. Xu, L. Cheng, X. Ma, G.-K. Chang, Digital mobile fronthaul based on delta-sigma modulation for 32 LTE carrier aggregation and FBMC signals. J. Opt. Commun. Netw. **9**(2), A233–A244 (2017)

41. B. Yu, C. Qian, P. Lin, J. Lee, Q. Li, S. Park, S. Kim, C. Yoon, S. Hu, L. Liu, Light-weight AI enabled non-linearity compensation leveraging high order modulations. IEEE Trans. Commun. **72**(1), 539–552 (2024)
42. J. Park, J. Lee, D. Kim, J.K. Choi, Deep reinforcement learning driven joint dynamic TDD and RRC connection management scheme in massive IoT networks. IEEE Access **12**, 34 973–34 992 (2024)
43. K. Boutiba, M. Bagaa, A. Ksentini, Multi-agent deep reinforcement learning to enable dynamic TDD in a multi-cell environment. IEEE Trans. Mob. Comput. **23**(5), 6163–6177 (2024)
44. Q. Tang, O. Ermis, C.D. Nguyen, A.D. Oliveira, A. Hirtzig, A systematic analysis of 5G networks with a focus on 5G core security. IEEE Access **10**, 18 298–18 319 (2022)
45. I.A. Alimi, R.K. Patel, A.O. Mufutau, N.J. Muga, A. Pinto, P.P. Monteiro, Towards a sustainable green design for next-generation networks. Wirel. Pers. Commun. **121**, 1123–1138 (2021). https://doi.org/10.1007/s11277-021-09062-2
46. I.A. Alimi, R.K. Patel, A. Zaouga, N.J. Muga, Q. Xin, A.N. Pinto, P.P. Monteiro, Trends in cloud computing paradigms: fundamental issues, recent advances, and research directions toward 6G fog networks, in *Moving Broadband Mobile Communications Forward*, ed. by A. Haidine (IntechOpen, Rijeka, 2021), ch. 1. https://doi.org/10.5772/intechopen.98315

5G FWA Customer Premise Equipment

5.1 Introduction

A key objective of 5G FWA is to provide performance levels comparable to wired and fiber technologies, including cable, xDSL, and FTTH, while substantially reducing costs [1–4]. While 4G FWA is a proven technology often employed to supplement current wired or fiber networks, 5G FWA, particularly in mid-band and high-band frequencies, is intended to serve as a replacement for traditional wired networks. Currently, 5G FWA offers typical data rates that can effortlessly surpass 500 Mbps, with peak speeds of up to 2000 Mbps. As CPEs evolve to support advanced features and higher bandwidths like multi-technology CA and multiband, the data rates for 5G FWA are expected to continue increasing [5].

5.2 FWA and Mobile CPEs

CPE encompasses a broad range of devices, including telephone handsets, modems, routers, and other equipment used within or outside a customer's premises, typically at their home or business, to link subscribers to either the public telecommunications network or the infrastructure network. CPE is essential for telecommunications services and infrastructure, significantly affecting the quality and reliability of service [6–8]. Using 5G CPE, the connection to the core network is established via the 5G RAN, utilizing either frequency range 1 (FR1) or FR2 frequencies [9–11].

Service providers use CPE to link fixed subscribers to the 5G RAN through a wireless connection. This approach is the quickest and most convenient way to deliver 5G to consumers in locations where fiber and cable connections are impractical or cost-prohibitive, including both dense urban and rural areas environments. The CPE receives the 5G signal from the BS and can also support mobile users through Wi-Fi or cellular methods. In this context, a 5G CPE device functions like a 5G router. It converts the 5G network signal into a Wi-Fi signal for transmission, and the data received via the Wi-Fi network is then converted back into a 5G signal for upload. This allows us to enjoy the high-speed benefits of 5G networks on our smartphones and laptops, even if they do not directly support 5G [9].

FWA CPE, built on the 3GPP standard, delivers high-speed Internet service. MNOs can easily deploy FWA within their existing mobile networks to deliver high-speed Internet, bypassing the need for the cable infrastructure usually demanded by traditional fixed broadband providers. FWA CPE supports both SA and NSA modes, offering flexibility for smooth integration with various 5G network configurations. Additionally, it operates in both mmWave and Sub-6 GHz frequency bands, guaranteeing compatibility with a range of 5G network bands and environments. The FWA CPE features flexible antenna designs suitable for a range of applications, including directional, omni-directional, and hybrid antennas. These alternatives allow users to tailor their connectivity solutions to meet specific needs. The FWA CPE also includes efficient device control and monitoring capabilities through TR-069, TR-181, and TR-369 protocols [12, 13].

Historically, the performance of CPE has often been a constraint on the general performance of mobile network links. However, because of the nature of FWA, CPEs tailored explicitly for FWA can leverage more relaxed requirements to address certain limitations associated with mobile CPEs [14]. This can lead to considerably larger cell sizes, along with higher peak and average data rates in comparison to those achieved with mobile CPEs. The following section explores the key differences [5].

5.2.1 Physical Size

Mobile or handheld CPEs are designed to be compact enough to fit in a pocket. In contrast, FWA CPEs can be significantly larger while still adhering to aesthetic standards that make them suitable for mounting in homes or on building exteriors [1, 5].

5.2.2 TX Power/EIRP Limitations

Mobile or handheld CPEs often face limitations on transmitter power due to Maximum Permissible Exposure (MPE) regulations, given their close proximity to users. In contrast, FWA CPEs, particularly outdoor models, are not handheld and are typically placed in locations where people are not near the antennas. As a result, these devices are permitted to operate with higher RF power and Effective Isotropic Radiated Power (EIRP) [5, 8].

5.2.3 Antennas and Advanced Features

Mobile CPEs generally employ antennas with broad beamwidths, and low directivity and gain due to the unpredictable and constantly changing orientation relative to the BS. In contrast, FWA CPEs are stationary and oriented in a fixed direction, allowing them to leverage advanced performance features more effectively. These features include fixed beamforming, which significantly enhances UL quality and reliability, and multiple MIMO layers, which improve throughput and link quality. Unlike mobile CPEs, which require processing overhead to track their location in a beamforming system, FWA CPEs do not have this requirement. These advantages result in superior performance for FWA CPEs compared to mobile CPEs [5].

5.2.4 Battery Life Issues

Mobile and handheld CPEs often face trade-offs between battery life and high UL/DL performance. In contrast, FWA CPEs are typically powered by building or house electricity, allowing them to be designed for higher performance without the same constraints on power consumption [5].

5.3 CPE Solutions

Choosing the right CPE form factor is crucial for designing effective FWA solutions. While cost considerations for the CPE itself are important, it is equally essential to evaluate how network expenses and revenue impacts influence the overall profitability of the CPE selection [15]. Placement of the CPE also plays a significant role in an operator's FWA strategy. Typically, subscribers may use indoor 5G FWA CPE options, such as desktop and window-mounted devices. However, alternative solutions include outdoor, rooftop-mounted antennas and emerging technologies like high-power indoor self-mounted CPEs. The decision between *Indoor* and *Outdoor* CPE options will impact product adoption and economic outcomes for a successful FWA business [5, 8]. Table 5.1 outlines the various features of sub-6 GHz 5G FWA across different deployment scenarios.

5.3.1 Indoor CPE

Indoor unit denotes a device designed for placement within a home. This is a highly appropriate option for customers, as it only requires finding a location with abundant signal strength within the house, typically close to a window. Indoor CPEs are popular because homeowners

Table 5.1 Features of sub-6 GHz FWA CPEs [15]

Segment/performance	Indoor gateway		Flexible installation receiver		Outdoor receiver	
Attenuation loss	High		Mid–low		None	
CPE antenna gain (dBi)	3–6	7–11	3–6	7–11	10–14	15–18
CPE size	In built		Smartphone	Tablet	Laptop	
CPE mounting options	Desktop		Flexible (desktop, window, wall)		Wall/rooftop	
Typical installation	Self-install				Technician	
Typical areas	Urban–suburban				Suburban–rural	

or businesses often prefer them, or even mandated due to restrictions on exterior mounting. Consequently, an FWA business scheme utilizing indoor CPE is anticipated to achieve higher customer adoption in comparison to other options.

An essential consideration for indoor mobile network-enabled CPE is the frequency bands employed for FWA, since not all bands propagate excellently indoors. Higher frequencies, specifically mmWave, can experience significant penetration losses–up to 40 dB–which makes them less suitable for indoor CPE implementations. Therefore, an Indoor CPE scheme is better suited to low-band or mid-band FWA. Mid-band 5G shows great promise and has substantial capability to support the majority of residential FWA applications [5, 8, 15, 16].

5.3.2 Outdoor CPE

The outdoor CPE is designed to be fitted on the exterior of a home or building. These units typically feature high-gain, directional antennas and generally offer superior transmission and reception performance compared to indoor CPEs. An outdoor CPE is regarded as a *controlled* network element, as it is installed in a specific location and remains fixed there. The main advantage of deploying an outdoor CPE is its ability to overcome propagation challenges associated with higher frequency bands. Additionally, the increased antenna gain of outdoor CPEs can significantly extend cell range, improving indoor coverage for mmWave bands and enhancing range for FWA solutions in rural areas. However, most current outdoor CPEs require professional installation, making them less convenient for customers compared to indoor CPEs. Operators also need to carefully consider the economic implications, including the additional installation costs per customer [5, 8, 15].

5.3.3 Flexi CPE

The CPE ecosystem is evolving quickly, featuring increasingly advanced designs aimed at overcoming the challenges of outdoor deployments. In this context, *window-mounted*

CPEs are becoming popular as a customer self-install option. They address the difficulties associated with professional outdoor CPE installations while benefiting from the enhanced link budget provided by external antennas. For instance, a new category of CPE, referred to as *flexible CPE*, seeks to merge the advantages of both indoor and outdoor CPEs. These versatile devices offer the convenience and quick deployment of indoor CPEs while achieving better spectral efficiency comparable to outdoor CPEs, such as higher antenna gains and reduced attenuation losses.

Flexible CPEs can be mounted either inside, near a window, or outside on a wall or roof. While some of these flexible CPEs provide similar antenna gains to indoor models, their performance is enhanced due to the elimination of attenuation losses. A major advantage of flexible CPEs is that they allow service providers to use the same unit type for homes both near and far from the radio site. The device must deliver the necessary antenna gain, power class, mounting options, and minimal attenuation loss to ensure ease of installation and optimal performance customer [5, 15].

5.4 5G NR CPE Power Classes

Power classes for mobile network-enabled CPEs are specified distinctly by the 3GPP for 5G NR in the FR1 (sub-6 GHz) and FR2 (mmWave) frequency ranges. For FR1, power classes are determined solely by maximum output RF power, excluding antenna gain. In contrast, FR2 power classes are characterized by both the maximum Total Radiated Power (TRP)–the total power emitted in all directions–and the minimum/maximum EIRP, which includes the power radiated in a specific direction along with antenna gain [13].

5.4.1 FR1 UE Power Classes

For FR1, 3GPP TS 38.101-1 defines UE power classes as detailed in Table 5.2. Class 2, in particular, is designed to support higher TRP for mid-band frequencies, that are frequently employed for FWA applications [5, 17].

Table 5.2 FR1 UE power classes [5]

Power classes	Max. RF output power (dBm)	Applicable bands
1	3	n14
1.5	29	n41
2	26	n41, n77, n78, n79
3	23	All other FR1 bands

Table 5.3 FR2 UE power classes [5]

Power classes	Max. TRP (dBm)	Min./Max. EIRP (dBm)	Notes
1	35	40/55	FWA UEs n260 min is 38 dBm
2	23	29/43	Automotive applications (radar, etc.)
3		22.4/43	Handheld UEs n260 min is 20.6 dBm
4		34/43	Non-handheld UEs n260 min is 31 dBm

5.4.2 FR2 UE Power Classes

For FR2, 3GPP TS 38.101-1 outlines UE power classes for FR2 bands n257, n258, n260, and n261, as detailed in Table 5.3. Power Class 1 is specifically designed for FWA UEs, allowing for substantially higher TRP and EIRP compared to other classes, due to safety considerations and previously established guidelines [5, 17–19].

5.5 FWA CPE Performance, Capabilities, and Applications

The primary distinction between indoor and outdoor CPE models lies in their capacity to meet expected service levels, particularly at peak periods. An indoor CPE device generally requires similar or slightly more radio resources compared to a smartphone, due to its indoor location. In contrast, an outdoor CPE device benefits from a 15–25 dB improvement in signal quality, resulting in lower cost per Mbps, higher speeds, and enhanced coverage. This advantage is especially significant for users located farther from the cell site, particularly in mid-band and mmWave deployments [15].

Several factors influence network gain when choosing CPE types, including antenna gain, attenuation, CPE power class, ISD, and spectrum frequency. An outdoor CPE delivers optimal performance due to its built-in directional antenna (e.g., 10–14 dBi at 3.5 GHz) and its installation with a predictable radio link quality to the chosen Radio BS. Typically, these devices are equipped with two receive antennas, though models equipped with four receive antennas are likewise offered. Additional receive antennas can be advantageous in urban settings, where multiple signal paths are accessible. Despite this, the transmission mode for a single CPE remains at rank-2, as the modem is expected to be installed with good LoS or near LoS [15].

A properly mounted outdoor CPE is aligned with the optimal serving cell, reducing link budget PL and enhancing the use of mid-band and mmWave TDD spectrum. The significant improvement in signal quality is attributed to the 10–15 dB increase in antenna gain and the elimination of 10–15 dB in wall/window attenuation losses experienced by indoor devices. Additionally, indoor devices face further signal attenuation due to their typical placement in

5.5 FWA CPE Performance, Capabilities, and Applications

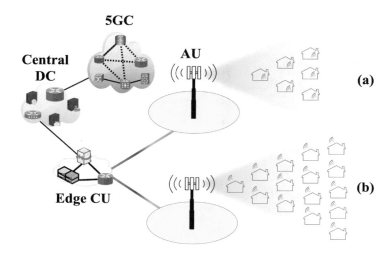

Fig. 5.1 Performance comparison of **a** indoor and **b** outdoor CPEs

the center of a home, away from windows, which can contribute an extra 5 dB in PL to ensure optimal Wi-Fi coverage. While indoor CPEs offer spectrum efficiency similar to that of smartphones, outdoor CPEs are typically two to three times more efficient. In practical terms, as illustrated in Fig. 5.1, this means that either two to three times more homes can be covered with the same amount of data consumption, or instead, two to three times more spectrum would be required to covered households with only indoor CPEs. Additionally, outdoor CPEs provide a more consistent performance across users, reducing the disparity between the worst, median, and best -performing users, which simplifies FWA commercial service agreements. Ensuring that users in weak signal conditions or at the cell edge have access to outdoor CPEs is crucial for maximizing the efficiency of radio network resources [15].

According to the study in [20], Fig. 5.2 illustrates the distribution of rooftop, wall-mounted, and indoor antennas for an ISD of 350 m. The findings indicate that, with an ISD of 350 m, 78% of households can utilize indoor antennas (often incorporated into CPEs). Meanwhile, the remaining households have to use either outdoor wall-mounted antennas (17%) or rooftop-mounted antennas (5%) to ensure improved signal propagation conditions.

End-users might experience adequate coverage and satisfactory speeds using an indoor device initially, but this situation may change rapidly as the cell becomes more congested with users. In a cell with no congestion, any user can be allocated numerous radio resources (PRBs) and attain satisfactory speeds, even if the radio link quality is subpar and just QPSK modulation is feasible. However, during peak hours when the cell is heavily loaded, the available resources are significantly reduced, and only devices capable of handling higher modulations (like 64-QAM or 256-QAM) will be able to maintain satisfactory speeds. The modulation and channel coding approach is guided by means of the devices' continuous

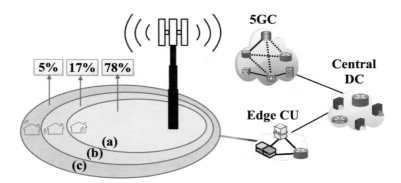

Fig. 5.2 Inter-site distance-based CPE breakdown for **a** indoor, **b** wall-mounted, and **c** rooftop antennas

Channel Quality Indicator (CQI) feedback in the UL. A subpar CQI value makes it costly for the radio network to support high data rates, such as for streaming video and TV, due to the need for lower modulation levels. Also, outdoor CPEs offer considerably better spectral efficiency compared to indoor CPEs. While outdoor CPEs are essential for homes in certain locations, such as those far from a BS, indoor CPEs can be adequate in areas close to the BS. Outdoor solutions tend to be more costly and complex to install, so there is a trade-off between achieving high performance and managing the costs of equipment and installation [15].

In cost-sensitive situations (such as those with lower ARPU) or where rapid deployment is essential, opting for indoor units is often practical, as they are generally less expensive and quicker to install. For instance, if there is ample available spectrum, an initial strategy might be to deploy indoor CPEs for most homes, with the possibility of upgrading to outdoor units later if necessary. Conversely, if the potential customer base in an area is large compared to the available spectrum, it may be beneficial to identify customers who would gain the most from outdoor CPEs during the sales process. In challenging propagation conditions, such as significant foliage, varied terrain, and obstacles obstructing LoS, a higher proportion of outdoor units may be necessary. Additionally, the enhanced spectral efficiency of outdoor CPEs typically extends range, thereby enlarging the potential customer base and increasing the proportion of outdoor CPEs [15, 21, 22].

5G NR technology offers versatile options for FWA services across various user types and environments. The optimal use cases are determined by the available spectrum, RF power, RF transmission characteristics, and other band-specific parameters. Table 5.4 provides overview of the peak throughput performance of 5G NR FWA CPEs across different bands, along with guidelines on the most likely applications for each band, considering both performance capabilities and economic factors [5].

Table 5.4 Analysis of FWA CPE performance and applications [5]

Frequency range (GHz)	Coverage	CPE peak throughput (Mbps)	Potential application
Low band (<2)	Excellent	Up to 250	Low-density rural, supplemental to other broadband technologies
Mid band (2.4–3.8)	Very good	Up to 1000	Medium-density suburban, can replace many existing wired broadband technologies
High band (>24)	Fair	Up to 2000	High-density urban, stadiums, malls, comparable speeds to fiber

Note These figures are expected to increase with the addition of new features in the future

5.6 Conclusion

Well-designed CPE is essential for providing excellent customer service and making the most of existing network setups. As a vital part of the FWA architecture, the strategic placement and advanced features of CPE can greatly enhance overall performance. Smaller, more visually appealing units not only simplify installation but also improve customer acceptance and satisfaction. Incorporating these factors into CPE design and deployment strategies can result in better service quality, increased customer retention, and more efficient network operation.

References

1. Y. Kimura, Y. Miura, T. Shirosaki, T. Taniguchi, Y. Kazama, J. Hirokawa, M. Ando, T. Shirouzu, A low-cost and very compact wireless terminal integrated on the back of a waveguide planar array for 26 GHz band fixed wireless access (FWA) systems. IEEE Trans. Antenn. Propag. **53**(8), 2456–2463 (2005)
2. B. De Beelde, E. Tanghe, D. Plets, W. Joseph, Outdoor channel modeling at D-band frequencies for future fixed wireless access applications. IEEE Wirel. Commun. Lett. **11**(11), 2355–2359 (2022)
3. M.K. Weldon, F. Zane, The economics of fiber to the home revisited. Bell Labs Tech. J. **8**(1), 181–206 (2003)
4. M. Jensen, R. Nielsen, O. Madsen, Comparison of cost for different coverage scenarios between copper and fiber access networks, in *2006 8th International Conference Advanced Communication Technology*, vol. 3 (2006), pp. 2015–2018
5. Fixed Wireless Access with 5G Networks, 5G Americas, White Paper (2021), https://www.5gamericas.org/wp-content/uploads/2021/11/5G-FWA-WP.pdf. Accessed 21 Apr 2024
6. W. Wan, W. Geyi, S. Gao, Optimum design of low-cost dual-mode beam-steerable arrays for customer-premises equipment applications. IEEE Access **6**, 16 092–16 098 (2018)

7. J. Proença, T. Cruz, P. Simões, M. Freitas, R. Calé, Virtualizing customer premises equipment, in *2021 IEEE Conference on Network Function Virtualization and Software Defined Networks (NFV-SDN)* (2021), pp. 104–105
8. I.A. Alimi, R.K. Patel, N.J. Muga, A.N. Pinto, A.L. Teixeira, P.P. Monteiro, Towards enhanced mobile broadband communications: a tutorial on enabling technologies, design considerations, and prospects of 5G and beyond fixed wireless access networks. Appl. Sci. **11**(21) (2021), https://www.mdpi.com/2076-3417/11/21/10427
9. Fixed Wireless Access: Bridging the connectivity gap with 5G FWA, Arctic Semiconductor, White Paper (2024), https://arcticsemiconductor.com/fixed-wireless-access/. Accessed 04 Jul 2024
10. What is Customer Premise Equipment (CPE)? Inseego Corp., White Paper (2024), https://inseego.com/resources/5g-glossary/what-is-cpe/. Accessed 04 Jul 2024
11. J. Du, D. Chizhik, R. Feick, G. Castro, M. Rodríguez, R.A. Valenzuela, Suburban residential building penetration loss at 28 GHz for fixed wireless access. IEEE Wirel. Commun. Lett. **7**(6), 890–893 (2018)
12. Broadband solutions: FWA CPE, Sercomm Corporation, Product information, https://www.sercomm.com/en/solutions/product/6/0/16. Accessed 1 Sept 2024
13. X. Tang, H. Chen, B. Yu, W. Che, Q. Xue, Bandwidth enhancement of a compact dual-polarized antenna for sub-6G 5G CPE. IEEE Antennas Wirel. Propag. Lett. **21**(10), 2015–2019 (2022)
14. R. Sun, D. Viorel, W. Keusgen, Outdoor-to-indoor loss measurement for rural/suburban residential scenario at 6 and 37 GHz, in *2023 17th European Conference on Antennas and Propagation (EuCAP)* (2023), pp. 1–5
15. Emerging CPE choices drive fast adoption, in *Ericsson, Fixed Wireless Access Handbook Extracted version, Insight 5 of 6* (2024), https://www.ericsson.com/4ade15/assets/local/reports-papers/further-insights/doc/fwa_insights_5_emerging_extracted.pdf. Accessed 05 Jul 2024
16. D. Schulz, J. Hohmann, P. Hellwig, C. Kottke, R. Freund, V. Jungnickel, R.-P. Braun, F. Geilhardt, All-indoor optical customer premises equipment for fixed wireless access. Opt. Fiber Commun. Conf. Exhib. (OFC) **2021**, 1–3 (2021)
17. M. Sharma, K. Sharma, S. Dwivedi, D.K. Singh, R. Gill, N. Kumar, A multiple millimeter-wave 5G MIMO antenna including n257/n258/n259/n260/n261 FR2-bands with high diversity performance, in *2023 International Conference on Power Energy, Environment & Intelligent Control (PEEIC)*, pp. 279–284 (2023)
18. F. Zhao, W. Deng, H. Jia, W. Ye, R. Wan, B. Chi, A band-shifting millimeter-wave T/R front-end with enhanced imaging and interference rejection covering 5G NR FR2 n257/n258/n259/n260/n261 bands, in *IEEE Radio Frequency Integrated Circuits Symposium (RFIC)*, pp. 29–32 (2023)
19. X. Zhang, J. Li, C. Yang, T. Liang, W. Zhang, Y. Yang, A wideband dual-polarized antenna-in-package for 5G millimeter-wave user equipment. IEEE Trans. Compon. Packag. Manuf. Technol. 1–1 (2024)
20. K. Laraqui, S. Tombaz, A. Furuskär, B. Skubic, A. Nazari, E. Trojer, Fixed wireless access on a massive scale with 5G. Ericsson, Technol. Rev. **94** (2017), https://www.ericsson.com/assets/local/publications/ericsson-technology-review/docs/2017/2017-01-volume-94-etr-magazine.pdf

References

21. J. Zhang, C. Masouros, Learning-based predictive transmitter-receiver beam alignment in millimeter wave fixed wireless access links. IEEE Trans. Signal Process. **69**, 3268–3282 (2021)
22. Z. El Khaled, W. Ajib, H. Mcheick, Log distance path loss model: application and improvement for sub 5 GHz rural fixed wireless networks. IEEE Access **10**, 52 020–52 029 (2022)

Wireless Network Coverage Planning

6.1 Introduction

With recent advancements in the telecommunications industry and the rollout of 5G networks, radio propagation modeling has become crucial for network planning and optimization [1]. Consequently, wireless network coverage planning is essential for MNOs and FWA providers to assess network performance and strategize future deployments of antenna masts and wireless access points (APs). Although 3G, 4G, and 5G employ different methods for sharing the wireless medium, all must efficiently manage their radio resources to ensure communication between users and their respective cells without interference from neighboring cells. This is challenging due to the scarcity of radio resources, which necessitates their reuse [2].

6.1.1 Fundamentals of Radio Propagation

In radio communication, a wireless signal is sent from a transmitter to a receiver through unguided free space. As the signal propagates, it can experience attenuation due to factors such as absorption, reflection, and refraction. The reduction in signal strength during propagation, known as PL, has been a significant focus of wireless research for decades. Understanding this loss is crucial because it supports various practical applications, including radio engineering, localization, and optimization of cellular systems. PL, typically measured in decibels (dB), is generally defined as the difference between the strength of the transmitted signal and the Received Signal Strength (RSS) [1, 3–5]. It is expressed as [6]

$$PL\,(dB) = 10\log\left(\frac{P_t}{P_r}\right), \qquad (6.1)$$

where P_t and P_r denote the transmitted and received power, respectively.

In addition to free space loss, signal strength can be affected by factors such as clutter height, terrain type, the characteristics of the propagation medium, the environment (rural, urban, or suburban), the height and number of buildings the signal must penetrate, the type of antenna deployed, the antenna's radiation pattern, and the placement of antennas, among others. PL calculation is essential for choosing BS locations, predicting coverage, calculating link budgets, and optimizing system performance. Directly measuring PL at the desired location is often costly and time-consuming. To mitigate these measurement costs, PL can be predicted instead. This process, known as *PL prediction*, estimates the amount of power that will be received at the receiving end, commonly referred to as *RSS prediction*, which is typically the primary objective of radio propagation models.

Radio propagation models are employed to estimate PL under different environmental conditions. So, they are essential tools in the telecommunications industry, aiding in the prediction of how waves propagate through diverse environments. They assist planners and engineers in projecting coverage, signal strength, and interference across different propagation scenarios, thereby supporting network planning and optimization [1] The models are typically designed for specific scenarios based on factors such as the distance between the transmitter and receiver, operating frequency, weather conditions, and other considerations. Underestimating or overestimating PL happens when an incorrect model is selected for a scenario, or when interference between cells is wrongly predicted. Thus, understanding PL in a specific geographical area enhances network planning [6]. These models are typically divided into two categories: Empirical models and Deterministic models [1].

6.1.1.1 Empirical Models

Empirical models, also known as statistical models, are based on observations and measurements. They rely on extensive experimental data and statistical analysis to determine the received signal level in a specific propagation medium. These models use the relationship between PL and environmental parameters for modeling and are often computationally efficient. Also, the model is more accurate and effective when deployed in environments similar to the one where it was developed. Therefore, the models are most accurate and suitable in the area where the measurement campaign was conducted, but they require adaptation for different areas. Numerous commercially available prediction tools are based on these models, including the Lee model, the Egli model, the Okumura-Hata model for urban and suburban environments, and the Cost 231 Walfisch-Ikegami model for dense urban environments. The latter is particularly useful in microcellular systems where antennas are deployed below the height of buildings [1, 6–11].

6.1.1.2 Deterministic Models

Deterministic models, also known as geometric models, estimate signal power directly from the path profile. These models utilize detailed environmental data, including 3D maps, terrain data, satellite images, and antenna types. Ray tracing is a widely used deterministic method

6.1 Introduction

for both indoor and urban scenarios [10]. The theory behind these models is derived by numerically solving Maxwell's equations. This approach adheres to the physical laws of electromagnetic wave propagation, utilizing diffraction and reflection principles to estimate the signal strength received at any given location. Deterministic models are deemed more reliable than empirical models because they account for a greater number of parameters in their calculations. However, this increased reliability comes at a cost, as deterministic models are more computationally demanding, requiring additional time and processing power. Deterministic models are typically employed for short propagation paths because the precision of the PL prediction increasingly relies on environmental details, such as shadowing or multipath effects, in these scenarios [1, 6]. Typically, radio signals attenuate as they travel through different transmission paths. Essentially, the ratio of received power to transmitted power, expressed in decibels (dB), with log-distance, accounts for the joint impacts of path loss, shadowing, and multipath fading. This relationship can be described as [12]

$$\frac{P_r}{P_t}(\text{dB}) = A + 10\alpha \log\left(\frac{d}{d_0}\right) + X_\sigma(\text{dB}) + Y(\text{dB}), \tag{6.2}$$

where A represents a constant influenced by the characteristics of the antenna and the attenuation of the channel, d_0 denotes the reference distance for the antenna's far field, α, represents the path loss exponent, X_σ and Y are random variables that account for shadowing and fast fading effects, respectively.

6.1.2 Link Planning Tools

Accurate and efficient radio propagation models facilitate the estimation of PL or RSS. These estimations are essential for various practical applications, such as creating radio coverage maps and performing localization [1]. Typically, terrain and surface elevation data, along with various propagation models, are used to create p2p wireless signal path profiles and evaluate RSS [13]. Coverage is assessed using these path profiles across different combinations of APs and customer locations. Several commercial tools, such as CellNetwork, Google Network Planner, cnHeat, ProgiraPlan, and CHIRplus_TC, are available for this purpose [14, 15]. These tools often require manual input of CPE antenna locations or involve calculating path profiles from APs to all geographical information system (GIS) pixels within the target area. The first method is labor-intensive and can become a bottleneck, while the second method demands substantial computational resources, especially for high spatial resolution, and requires regenerating coverage maps whenever APs are changed. To overcome these challenges, an AI-based universal enhancement for wireless coverage planning tools can be utilized.

6.2 AI-Based Link Planning Tools

The widespread use of wireless devices results in the generation of more data on radio propagation behavior. Additionally, advancements in measurement and instrumentation have made it quite easier to measure radio signals in various scenarios. The availability of diverse data, combined with the enhanced efficiency of ML computation, allows ML-based radio models to achieve greater accuracy. This improvement has led to a rise in the number of these models within the community. Common ML for predicting PL include Random Forest (RF), Decision Tree (DT), K-Nearest Neighbors (KNN), Support Vector Regression (SVR), Neural Network (NN), and Convolutional Neural Network (CNN) [1].

Furthermore, the proliferation of wireless devices has heightened the competition for radio spectrum resources and increased interference. As a result, there is now an expectation for radio signals to be generated more intelligently to minimize interference. The expectation has driven the wireless industry towards creating intelligent cellular architectures that dynamically adjust network element usage based on service demand and automate operations to minimize both CAPEX and OPEX. This entails creating efficient, unsupervised RAN planning, which directly influences system expenditures. The goal of this intelligent cellular planning is to determine BS configurations (such as user associations, coverage, and antenna radiation patterns) that minimize the number of deployed BSs while satisfying coverage and capacity requirements. The complexity and lack of scalability of the cellular planning optimization problem have been demonstrated. Additionally, most current cellular planning methods lead to either over-provisioned or under-provisioned architectures [16–19]. This has also spurred greater efforts to utilize ML models in radio transmissions [1]. This approach optimizes the use of computing resources and eliminates the need for manual input. Specifically, CPE antenna locations can be automatically extracted from aerial images of the target buildings [14].

6.2.1 Evolutionary Algorithms for Network Planning

In a heterogeneous network (HetNet) architecture, installing mmWave small cells alongside current macro-relay networks creates a mixed cell or hierarchical structure. This network can be planned by integrating sub-6 GHz spectrum Macro-cell RRUs (MRRUs) with mmWave Micro-cell RRUs (mRRUs). The mRRU antennas deliver high data rates but have a restricted coverage range, while the MRRU antennas offer extended range coverage [20–23]. To satisfy coverage and bandwidth needs, mRRUs and MRRUs need to be strategically placed. This involves a hybrid approach where both the quantity and placement of MRRUs and mRRUs are optimized to meet user traffic demands and coverage constraints. To achieve this, system parameters and user requirements for the target area need to be defined. Initial coverage and cell capacity planning phases will determine the number of MRRUs and mRRUs required to meet the targeted QoS. This is followed by a joint placement phase to optimize the locations of MRRUs and mRRUs, ensuring that coverage and the desired DL and UL data rates are achieved [24].

Meanwhile, Particle Swarm Optimization (PSO) has evolved into a more advanced optimization algorithm by continuously iterating to search for the optimal solution and utilizing the value of the objective function. So, evolutionary algorithms, such as Simulated Annealing and Accelerated PSO (A-PSO), can be employed to independently identify the optimal locations for MRRUs and mRRUs. Moreover, an iterative method can be used to remove redundant MRRUs and mRRUs while ensuring the necessary coverage and data rates for each subarea [24, 25].

6.2.2 Machine Learning-Based 3D RF Planning Tool

A customized 3D RF planning tool, utilizing precise map data, sophisticated radio propagation models, and high-speed Graphical Processing Units (GPUs), can significantly enhance network planning processes. This tool accurately simulates mmWave beamforming coverage in a given area, producing results in minutes to hours. The tool features a Semantic Map that integrates map data from diverse sources, such as satellite imagery, 3D building models, and street or candid photography capturing unfiltered chance encounters and unexpected moments in public spaces. Using ML techniques, the RF planning tool develops a precise internal model of a neighborhood, incorporating the locations and geometries of trees, buildings, windows, and street furniture. With the Semantic Map, potential BS sites can be identified, and ray tracing is then employed to model RF beam patterns from these sites across the environment. Ray tracing calculates the trajectories of numerous lines as they interact with various objects in the environment, a process that is computationally intensive but well-suited to modern computer graphics cards. So, using the 3D RF planning tool, an entire neighborhood can be accurately simulated within minutes, allowing for the selection of optimal BS sites. Experts can then make further adjustments within the tool to fine-tune the FWA deployment according to specific service requirements. This enables providers to utilize mmWave radio access effectively for delivering FWA services [26].

6.2.3 AI-Based Computer Vision Aided Network Coverage Planning

Wireless signal propagation is highly dependent on the placement and heights of both the CPE and AP antennas, in addition to the obstructions along the signal propagation path. This sensitivity increases with higher frequencies and narrower antenna beams. Even minor adjustments, such as moving the CPE antenna a few meters, can significantly alter the signal propagation environment. For instance, relocating the antenna from one side of a pitched roof to the other can shift the signal from a LoS path to a NLoS path, which can greatly impact the QoS [14, 27–29].

To precisely assess signal coverage at subscriber locations, a coverage planning tool must identify potential CPE antenna locations where adequate RSS is expected. Given the safety considerations and time constraints faced by field engineers, roof edges are commonly chosen for mounting antennas, as they are easily accessible and minimize the risk of breaking tiles while maintaining elevation. Another typical installation location is at the top of a chimney, which often provides the peak height on a structure, though chimneys are less common on modern buildings. AI-based computer vision methods can be employed to detect roof edges as possible antenna installation locations. The coordinates can then be provided to coverage planning tools to generate comprehensive signal path profiles [14].

AI has seen rapid advancements, with Deep Learning, a major subsection of AI, becoming widely utilized in real-world applications like speech recognition, object classification/detection, and language translation [30]. The advancements in Neural Networks (NN)-based techniques, such as Convolutional Neural Networks (CNNs) and Deep Neural Networks (DNNs), have led to the development of robust tools for image processing operations that are challenging for conventional computer vision approaches [31]. The effectiveness of object detection has notably improved with the introduction of Region-based CNN (R-CNN), and its subsequent variants, Faster R-CNN and Mask R-CNN, have advanced semantic and instance segmentation for images and real-time videos [32–35]. Large image datasets like Microsoft COCO, PASCAL VOC, and ImageNet facilitate the training of DNNs to recognize common objects [36–41]. Pretrained DNN models such as ResNets, AlexNets, and GoogLeNets enable developers to leverage these models as backbones and quickly adapt them for various object detection and classification operations [42–45]. For instance, Faster R-CNN with a ResNet-101 backbone can be employed to identify industrial chimneys in remote sensing images [14, 46, 47].

Different types of CNNs are extensively utilized for building extraction and segmentation from satellite and aerial images. Additionally, various Fully Convolutional Network (FCN) models can be used to extract building footprints. For buildings with irregular shapes, Faster Edge Region CNN (FER-CNN) can enhance detection results. Similarly, Framed Field Learning has been applied to UNet architectures to address the challenges of predicting footprints for buildings with non-standard shapes [14, 48–51].

6.3 Conclusion

Traditional wireless network coverage planning is a complex and resource-intensive task, often demanding significant time, manpower, and costly equipment to achieve comprehensive coverage and prevent gaps. Inadequate planning can lead to poor coverage quality or even complete black spots, resulting in high costs and extended resolution times. However, integrating AI into coverage planning tools has been a game-changer. AI-based methods can enhance test points, making the planning process more accurate and efficient, and significantly improving the utility of coverage tools. This approach not only streamlines planning but also facilitates more precise CPE installation, leading to better network performance and increased customer satisfaction.

References

1. M. Vasudevan, M. Yuksel, Machine learning for radio propagation modeling: a comprehensive survey. IEEE Open J. Commun. Soc. **5**, 5123–5153 (2024)
2. R. Cortesão, D.F.S. Fernandes, G.E. Soares, D.J.A. Clemente, P.J.A. Sebastião, L.S. Ferreira, Cloud-based implementation of a SON radio resources planning system for mobile networks and integration in SaaS metric. IEEE Access **9**, 86 331–86 345 (2021)
3. J.J. Popoola, D. Okhueleigbe, I.A. Alimi, Link adaptation for microwave link using both MATLAB and path-loss tool. Indones. J. Electr. Eng. Inf. **4**, 281–291 (2016), https://section.iaesonline.com/index.php/IJEEI/article/viewFile/245/168
4. I.A. Alimi, J.J. Popoola, K.F. Akingbade, A power efficient rake receiver for interference reduction in the mobile communication systems. Int. J. Electron. Electr. Eng. **3**(6), 1–5 (2016)
5. I.A. Alimi , O. Aboderin, Adaptive interference reduction in the mobile communication systems. Am. J. Mob. Syst. Appl. Serv. **1**(1), 1–8 (2015), http://www.aiscience.org/journal/paperInfo/ajmsas?paperId=1153
6. O.O. Erunkulu, A.M. Zungeru, C.K. Lebekwe, J.M. Chuma, Cellular communications coverage prediction techniques: a survey and comparison. IEEE Access **8**, 113 052–113 077 (2020)
7. A.D.S. Braga, H.A.O.D. Cruz, L.E.C. Eras, J.P.L. Araújo, M.C.A. Neto, D.K.N. Silva, G.P.S. Cavalcante, Radio propagation models based on machine learning using geometric parameters for a mixed city-river path. IEEE Access **8**, 146 395–146 407 (2020)
8. R. Borralho, A.U. Quddus, A. Mohamed, P. Vieira, R. Tafazolli, Coverage and data rate analysis for a novel cell-sweeping-based RAN deployment. IEEE Trans. Wirel. Commun. **23**(1), 217–230 (2024)
9. T. Rao, S. Rao, M. Prasad, M. Sain, A. Iqbal, D. Lakshmi, Mobile radio propagation path loss studies at VHF/UHF bands in Southern India. IEEE Trans. Broadcast. **46**(2), 158–164 (2000)
10. D. Erricolo, P. Uslenghi, Propagation path loss—a comparison between ray-tracing approach and empirical models. IEEE Trans. Antennas Propag. **50**(5), 766–768 (2002)
11. K.L. Chee, S.A. Torrico, T. Kurner, Radiowave propagation prediction in vegetated residential environments. IEEE Trans. Veh. Technol. **62**(2), 486–499 (2013)
12. R. He, Z. Zhong, B. Ai, J. Ding, K. Guan, Analysis of the relation between fresnel zone and path loss exponent based on two-ray model. IEEE Antennas Wirel. Propag. Lett. **11**, 208–211 (2012)
13. S.I. Popoola, A.A. Atayero, N. Faruk, Received signal strength and local terrain profile data for radio network planning and optimization at GSM frequency bands. Data Brief **16**, 972–981 (2018), https://www.sciencedirect.com/science/article/pii/S235234091730731X
14. Y. Chu, H. Ahmadi, D. Grace, D. Burns, Deep learning assisted fixed wireless access network coverage planning. IEEE Access **9**,124 530–124 540 (2021)
15. S.-K. Ahn, S. Ahn, J. Kim, H. Kim, S. Kwon, S. Jeon, M. Ek, S. Simha, A. Saha, P.M. Maru, P. Naik, M. Aitken, P. Angueira, Y. Wu, S.-I. Park, Evaluation of ATSC 3.0 and 3GPP Rel-17 5G broadcasting systems for mobile handheld applications. IEEE Trans. Broadcast. **69**(2), 338–356 (2023)
16. M. Chraiti, A. Ghrayeb, C. Assi, N. Bouguila, R.A. Valenzuela, A framework for unsupervised planning of cellular networks using statistical machine learning. IEEE Trans. Commun. **68**(5), 3213–3228 (2020)

17. A. Engels, M. Reyer, X. Xu, R. Mathar, J. Zhang, H. Zhuang, Autonomous self-optimization of coverage and capacity in LTE cellular networks. IEEE Trans. Veh. Technol. **62**(5), 1989–2004 (2013)
18. C.K. Anjinappa, S. Güvenç, Coverage hole detection for mmWave networks: an unsupervised learning approach. IEEE Commun. Lett. **25**(11), 3580–3584 (2021)
19. P.M. Pina, A.F. Godinho, D.F.S. Fernandes, D.J.A. Clemente, P. Sebastião, G.E. Soares, L.S. Ferreira, Automatic coverage based neighbour estimation system: a cloud-based implementation. IEEE Access **8**, 69 671–69 682 (2020)
20. A.K. Saurabh, M.K. Meshram, Integration of sub-6 GHz and mm-Wave antenna for higher-order 5G-MIMO system. IEEE Trans. Circ. Syst. II: Express Briefs **69**(12), 4834–4838 (2022)
21. Z. Gu, H. Lu, P. Hong, Y. Zhang, Reliability enhancement for VR delivery in mobile-edge empowered dual-connectivity sub-6 GHz and mmWave HetNets. IEEE Trans. Wirel. Commun. **21**(4), 2210–2226 (2022)
22. L. Yashvanth, C.R. Murthy, On the impact of an irs on the out-of-band performance in sub-6 GHz & mmWave frequencies. IEEE Trans. Commun. 1–1 (2024)
23. X. Xia, F. Wu, C. Yu, Z. Jiang, J. Xu, S.-Y. Tang, Z. Wang, Y. Yao, W. Hong, Millimeter-wave and sub-6-GHz aperture-shared antenna and array for mobile terminals accessing 5G/6G-enabled IoT scenarios. IEEE Internet of Things J. **11**(10), 18 808–18 823 (2024)
24. H. Ganame, L. Yingzhuang, A. Hamrouni, H. Ghazzai, H. Chen, Evolutionary algorithms for 5G multi-tier radio access network planning. IEEE Access **9**, 30 386–30 403 (2021)
25. X. Xu, J. Li, M. Zhou, J. Xu, J. Cao, Accelerated two-stage particle swarm optimization for clustering not-well-separated data. IEEE Trans. Syst. Man, Cybern. Syst. **50**(11), 4212–4223 (2020)
26. 5G Fixed Wireless Access, Samsung, White Paper (2018), https://images.samsung.com/is/content/samsung/p5/global/business/networks/insights/white-paper/samsung-5g-fwa/white-paper_samsung-5g-fixed-wireless-access.pdf. Accessed 05 Jul 2024
27. Y. Rao, Z. hui Jiang, N. Lazarovitch, Investigating signal propagation and strength distribution characteristics of wireless sensor networks in date palm orchards. Comput. Electron. Agric. **124**, 107–120 (2016), https://www.sciencedirect.com/science/article/pii/S0168169916300928
28. F. B. Teixeira, R. Campos, M. Ricardo, Height optimization in aerial networks for enhanced broadband communications at sea. IEEE Access **8**, 28 311–28 323 (2020)
29. Y. Yu, Y. Liu, W.-J. Lu, H.-B. Zhu, Path loss model with antenna height dependency under indoor stair environment. Int. J. Antennas Propag. **2014**(1), 482615 (2014), https://onlinelibrary.wiley.com/doi/abs/10.1155/2014/482615
30. L. Deng, D. Yu (2014)
31. V. Sze, Y.-H. Chen, T.-J. Yang, J.S. Emer, Efficient processing of deep neural networks: a tutorial and survey. Proceed. IEEE **105**(12), 2295–2329 (2017)
32. R. Girshick, J. Donahue, T. Darrell, J. Malik, Rich feature hierarchies for accurate object detection and semantic segmentation, in *IEEE Conference on Computer Vision and Pattern Recognition* (2014), pp. 580–587
33. K. Elgazzar, S. Mostafi, R. Dennis, Y. Osman, Quantitative analysis of deep learning-based object detection models. IEEE Access **12**, 70 025–70 044 (2024)
34. Y.R. Musunuri, O.-S. Kwon, S.-Y. Kung, Cross-transfer learning for enhancing object detection in remote sensing images. IEEE Geosci. Remote Sens. Lett. **21**, 1–5 (2024)
35. R. Kaur, S. Singh, A comprehensive review of object detection with deep learning. Digit. Signal Process. **132**, 103812 (2023), https://www.sciencedirect.com/science/article/pii/S1051200422004298
36. H. Wu, W. Gao, X. Xu, Solder joint recognition using mask R-CNN method. IEEE Trans. Compon. Packag. Manuf. Technol. **10**(3), 525–530 (2020)

37. S. Ren, K. He, R. Girshick, J. Sun, Faster R-CNN: towards real-time object detection with region proposal networks. IEEE Trans. Pattern Anal. Mach. Intell. **39**(6), 1137–1149 (2017)
38. X. Zhu, J. Li, J. Cao, D. Tang, J. Liu, B. Liu, Semantic-guided representation enhancement for multi-label image classification. IEEE Trans. Circ. Syst. Video Technol. 1–1 (2024)
39. S. Cai, L. Li, X. Han, S. Huang, Q. Tian, Q. Huang, Semantic and correlation disentangled graph convolutions for multilabel image recognition. IEEE Trans. Neural Netw. Learn. Syst. 1–13 (2023)
40. Y. Li, C. Wu, L. Li, Y. Liu, J. Zhu, Caption generation from road images for traffic scene modeling. IEEE Trans. Intell. Transp. Syst. **23**(7), 7805–7816 (2022)
41. H. Xu, X. Lv, X. Wang, Z. Ren, N. Bodla, R. Chellappa, Deep regionlets: blended representation and deep learning for generic object detection. IEEE Trans. Pattern Anal. Mach. Intell. **43**(6), 1914–1927 (2021)
42. W. Zheng, Y. Ai, W. Zhang, Hybrid improved concave matching algorithm and ResNet image recognition model. IEEE Access **12**, 39 847–39 861 (2024)
43. G. Li, Q. Liu, Z. Guo, Driver distraction detection using advanced deep learning technologies based on images. IEEE J. Radio Freq. Ident. **6**, 825–831 (2022)
44. P. Su, Y. Liu, S. Tarkoma, A. Rebeiro-Hargrave, T. Petäjä, M. Kulmala, P. Pellikka, Retrieval of multiple atmospheric environmental parameters from images with deep learning. IEEE Geosci. Remote Sens. Lett. **19**, 1–5 (2022)
45. P. Lyu, J. Liu, Y. Zhang, B. Ye, T. Lan, L.-P. Bai, Z. Cai, Z.-H. Jiang, A novel feature fusion framework for industrial automation single-multiple object detection. IEEE Trans. Ind. Inf. **20**(5), 7686–7697 (2024)
46. C. Han, G. Li, Y. Ding, F. Yan, L. Bai, Chimney detection based on faster R-CNN and spatial analysis methods in high resolution remote sensing images. Sensors **20**(16) (2020), https://www.mdpi.com/1424-8220/20/16/4353
47. M. Lalak, D. Wierzbicki, Methodology of detection and classification of selected aviation obstacles based on UAV dense image matching. IEEE J. Select. Top. Appl. Earth Obs. Remote Sens. **15**, 1869–1883 (2022)
48. D. Marzi, J.I.S. Jara, P. Gamba, A 3-D fully convolutional network approach for land cover mapping using multitemporal sentinel-1 SAR data. IEEE Geosci. Remote Sens. Lett. **21**, 1–5 (2024)
49. H. Cui, Y. Wang, Y. Li, D. Xu, L. Jiang, Y. Xia, Y. Zhang, An improved combination of faster R-CNN and U-Net network for accurate multi-modality whole heart segmentation. IEEE J. Biomed. Health Inform. **27**(7), 3408–3419 (2023)
50. L. Lv, Y. Zhao, X. Li, J. Yu, M. Song, L. Huang, F. Mao, H. Du, UAV-based intelligent detection of individual trees in Moso bamboo forests with complex canopy structure. IEEE J. Select. Top. Appl. Earth Obs. Remote Sens. **17**, 11 915–11 930 (2024)
51. N. Bhavana, M.M. Kodabagi, B.M. Kumar, P. Ajay, N. Muthukumaran, A. Ahilan, POT-YOLO: real-time road potholes detection using edge segmentation-based yolo V8 network. IEEE Sens. J. **24**(15), 24 802–24 809 (2024)

Key Methods for Efficient and High-Speed FWA Solutions

7.1 Introduction

The cost, time, and complexity associated with delivering fixed broadband have historically posed significant challenges to the deployment of high-speed data services. Previous efforts have often failed primarily because of the high expenses related to new infrastructure and equipment [1]. Additionally, the current mobile communication bands are insufficient to handle future traffic demands, necessitating the use of new frequency bands to provide larger spectrum blocks for operators. Accessing these wider bands requires utilizing higher frequencies, which presents its own set of challenges. Higher frequencies face a more challenging link budget due to increased free-space PL, as well as extra losses from diffraction and penetration. For example, 5G mmWave and 5G NR operations in sub-THz communications struggle to offer long-distance wireless transmission [2–4].

To effectively use high-frequency bands and deliver broadband applications requiring high traffic capacity and extended transmission range, new and improved technical solutions are required [2, 3]. One promising approach is the extended-range 5G FWA, which can utilize existing infrastructure such as macro sites and radio towers, leveraging common components to deliver high-speed broadband services to previously underserved consumer and enterprise markets [1].

Additionally, there are innovative methods for utilizing mmWave technology for backhaul, fronthaul, and air interface connectivity. For instance, IAB, with its various configurations and antenna deployment options, can reduce the number of BS sites needing fiber backhaul, enabling faster deployments and lower costs. Furthermore, repeaters, which have traditionally been used to extend signal coverage in Wi-Fi and 2G/3G/4G networks, are now evolving. Modern smart repeaters can process side control information and employ phased array antennas to more intelligently serve UE [5].

7.2 Millimeter-Wave and Enabling FWA Solutions

With the standardization of 5G NR technology now complete, MNOs are beginning to implement these systems globally. However, the rollout of 5G in the mmWave bands, which are essential for achieving the high bitrates specified by ITU for IMT-2020 systems, faces significant challenges. Issues such as LoS blockage and micromobility result in recurrent outages and diminished service quality [6]. This section explores strategies to overcome these challenges and enhance the utilization of mmWave spectrum for FWA.

7.2.1 MmWave-Only SA

A new network option, mmWave-only SA, supports the deployment and operation of NR exclusively on mmWave frequencies, without the need for a sub-6 GHz anchor band. This configuration enables service providers to offer gigabit FWA using only mmWave spectrum. Figure 7.1 depicts the mmWave-only SA solution. It is essential to recognize that mmWave-only FWA deployments require careful consideration of coverage and reliability, especially when compared to combined mid-band and mmWave FWA solutions. The inclusion of mid-band spectrum enhances reliability by providing a fallback option during mmWave outages caused by blockage or fading.

7.2.2 Extended-Range MmWave 5G FWA

The growing reliance on HBB and the increasing digitalization of homes and businesses are expected to boost FWA traffic volumes in the future. As the average revenue per user for operators is unlikely to grow proportionally, it is essential to develop FWA solutions with cost efficiency in mind [1, 7].

Mid-bands utilizing TDD and FDD low bands are often adequate for many FWA scenarios. Specifically, 3GPP TDD mid-bands like n41 (2.5–2.7 GHz) and n77 (3.3–4.2 GHz) offer significant opportunities for combining MBB with FWA services [7].

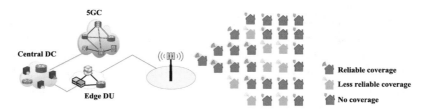

Fig. 7.1 mmWave-only FWA deployment

7.2 Millimeter-Wave and Enabling FWA Solutions

In dense suburban areas, high-end services frequently require additional capacity. To address this, high-band mmWave spectrum, like 26 and 28 GHz, can be utilized. This can be implemented either at existing macro sites or by adding new street-level sites for densification, providing exceptionally high peak rates, low latency, and substantial capacity. Although mmWave spectrum is typically linked to dense deployments [7, 8], its unfavorable propagation characteristics result in a much smaller cell coverage area compared to lower frequencies. At present, mmWave cells are usually designed to cover distances of up to approximately 600 m, with an ISD of almost 1 km [9].

In FWA deployments where there are favorable LoS conditions between outdoor rooftop-mounted CPEs and high tower radio sites, mmWave cell ranges can extend to multiple kilometers. Achieving this demands high-power CPEs, high-power radios, and advancements in software, specifically mmWave extended range technology, to manage the increased propagation time over longer distances [9]. This capability offers operators a significant prospect to broaden the deployment of mmWave spectrum for FWA in sparsely populated suburban and semi-rural regions [7, 8].

The 3GPP NR specification supports extended cell ranges for mmWave by incorporating a guard period in the TDD format. This guard period, which is the interval between DL and UL transmissions, accommodates the roundtrip time for signals and allows equipment to switch between reception and transmission. Although extending the cell range with a larger guard period can decrease DL peak rates for all users, it enables mmWave FWA to effectively serve certain households located several kilometers away, as long as they are in favorable signal conditions [9].

The farther a household is from a radio site, the more difficult it becomes to maintain excellent LoS conditions. An effective strategy involves combining mmWave and mid-band spectrum: mmWave can serve households in optimal conditions, while mid-band covers those in less favorable conditions. This approach allows mmWave to offload capacity, enabling mid-band to serve even more households at greater distances and in challenging locations. Consequently, operators can extend mmWave FWA services to more sparse suburbs and semi-rural areas [7, 8]. This strategy not only facilitates high-end wireless fiber services in more rural regions but also increases the number of households served per radio site and supports higher data consumption. It represents a significant opportunity to maximize the business and technological potential of mmWave FWA while leveraging mid-band spectrum [9].

The concept is demonstrated in Fig. 7.2, where a macro cell site, spanning some kilometers, is fitted with 5G NR radios for both mmWave and mid-band frequencies. In this setup, the system utilizes all available spectrum assets for FWA services across the sector. Homes within mmWave coverage are served primarily by mmWave, while those outside mmWave coverage are provided by mid-band. The extended range of mmWave in these implementations is achieved via a novel enhancement called mmWave extended range [7]. This functionality significantly extends coverage, increasing the effective range from the typical 600–900 m to over 5 km [1, 10].

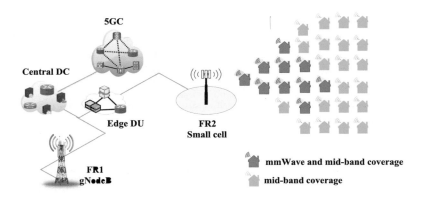

Fig. 7.2 mmWave extended range implementation

The extended range also expands the area where households can access mmWave services by approximately 40 times. This development is currently being rolled out commercially in several cities and represents a significant milestone, redefining the value of mmWave spectrum. It enables providers' mmWave home Internet solutions to deliver speeds exceeding 300 Mbps, which is 10–15 times faster than the speeds offered by their 4G LTE home Internet services [1, 10].

From a technical standpoint, mmWave extended-range technology relies on LoS to achieve its optimal performance. Households that cannot establish a direct LoS between the tower and their location can still be served using mid-band spectrum, ensuring that a broad range of consumers benefits from the service. This illustrates the complementary nature of mid-band and mmWave spectrums. In suburban and rural areas, where broadband options are often limited and providers may be scarce, wireless technologies become crucial. As distances between households increase, so do the costs of installing wired technologies. Wireless solutions can help address these challenges by leveraging existing infrastructure, thus keeping costs manageable for both providers and users. While mmWave technology was previously used in urban areas where range was less of an issue, relying on street furniture or micro towers for a 500 m coverage, this innovation extends the mmWave range, allowing more previously underserved households to benefit from high-speed broadband services [1].

Another advantage of the extended-range mmWave solution is its rapid deployment. Leveraging existing macro towers that are already serving mobile broadband customers, upgrading these towers and installing CPE for mmWave signals can be done much more quickly and at a lower initial cost compared to alternatives like fiber. Moreover, the development of tools and applications to assist installers in setting up the CPE has reduced the number of visits needed to adjust the equipment at consumers' premises. This not only leads to additional cost savings but also minimizes potential disruptions for consumers [1].

7.2 Millimeter-Wave and Enabling FWA Solutions

To effectively manage an FWA network in the mmWave spectrum over extended distances, two key aspects must be addressed: managing the extended propagation delay and optimizing the received signal strength [7].

7.2.2.1 Signal Strength Maximization

Because of the short wavelengths of mmWave signals, the antenna elements are also small, which means that the signal at the antenna needs to have a high energy density to achieve an adequate signal-to-noise ratio. To focus energy effectively in the desired direction, huge antenna arrays on the transmitter end, coupled with beamforming, are utilized. Also, beamforming on the receiver end additionally enhances the received signal strength. In general, beamforming is essential for transmission at mmWave frequencies and can be implemented on both the device and network sides, though network-side beamforming tends to be more sophisticated. However, beamforming by itself is insufficient to achieve spans of several kilometers [7]. Current mobile networks utilizing mmWave frequencies employ beamforming yet still fail to connect subscribers located at considerable distances.

In FWA, it is feasible to establish conditions that favor extended-distance coverage. On the network side, high-power radios installed beyond major obstructions, such as those mounted on macro towers, are well-positioned to optimize DL signal strength. On the device side, AC-powered CPEs offer considerably increased transmit power compared to battery-operated mobile devices. Additionally, they can be positioned in optimal locations, typically outdoors, to minimize wall penetration loss. Both the high transmit power of the CPE and its strategic placement are crucial for enhancing UL signal strength, which often constrains coverage [7].

7.2.2.2 Addressing Prolonged Delay

Accommodating long propagation delays is crucial for effective long-distance communication, and the extended range characteristic facilitates this. After adequate signal strength is achieved at a specific range, the communication system must be adjusted to handle the corresponding propagation delay. In mmWave spectrum, TDD is used, which involves alternating between DL and UL slots. The transceivers need a small number of microseconds to switch between receiving and transmitting modes, necessitating a brief gap between UL and DL symbols in the TDD format. The duration of this gap must account for the time it takes for the signal to travel from the transmitter to the receiver. The greater the range between them, the larger the gap required. In deployments focused on MBB, where the range of mmWave cells usually spans only a few hundred meters, the mmWave transmissions require only a brief gap between the DL and UL slots in a TDD pattern. For FWA deployments designed for longer distances, the gap between DL and UL needs to be extended. This involves muting additional data symbols to account for the increased distance. Along with a longer gap period, a 3GPP-defined random access preamble format optimized for long-distance scenarios is utilized. Random access preambles, integral to the 3GPP Release 15 specifications, are compatible with all devices. On the network side, detecting these preambles, which are affected by significant propagation delays, can be enhanced using an innovative random access receiver algorithm [7].

The increased gap period in the TDD pattern necessary for covering longer distances adds some overhead for all devices in larger cells, resulting in a slight reduction in peak rates and capacity. Consequently, the extended range setup must be utilized just when absolutely necessary [7].

7.3 Advancement of Network Nodes

As we transition to higher operating frequencies, such as mmWave in the NR standard and the THz frequency range for 6G, network coverage challenges will intensify. To address these coverage issues, various network nodes, including IAB, repeaters, and intelligent reflective surfaces (IRS), have been introduced. These nodes not only extend coverage but also boost end-user throughput [11–13].

7.3.1 Integrated Access and Backhaul

5G NR technology lays the groundwork for future wireless and mobile communications by accommodating novel application categories and flexible spectrum use [14]. In a 5G network, transport connectivity can be achieved by either wireline or wireless methods. While fiber optics are the top option for connectivity [15], the densely packed small cell structure of 5G networks render it expensive and labor-intensive for MNOs to offer fiber backhaul to each AP [14]. This issue also affects cell-free massive MIMO networks, where numerous distributed APs require backhaul links [16].

As networks become denser, particularly with the use of mmWave spectrum in the RAN, wireless solutions become increasingly viable [15]. Wireless backhaul offers a promising alternative for dense future deployments due to its flexibility, cost-effectiveness, high spectrum efficiency, and mobility benefits [16]. It is also quicker and simpler to deploy and reconfigure in comparison to wired backhaul [17]. IAB stands out as a critical technology for 5G NR, addressing these challenges by using the significant amounts of spectrum available in mmWave frequencies to allow wireless spectrum sharing between access and backhaul [12, 14, 18, 19].

With extremely wide bandwidths, it becomes increasingly relevant to explore the feasibility of incorporating wireless backhaul transport within the same spectrum as wireless access. This can be approached in two ways: by either allocating a fixed portion of the spectrum for backhaul and another for access traffic, or by dynamically adapting the spectrum allocation according to real-time requirements. While dynamic allocation can lead to more efficient spectrum utilization, it also introduces greater system intricacy. In such integrated systems,

7.3 Advancement of Network Nodes

where backhaul and access share the same technology, both the radio resources and interface, such as hardware, are utilized for both backhaul and access links. An example of this is in-band relaying in 3GPP LTE Release 10, where a Donor evolved Node B (DeNB) offers backhaul connections to a Relay Node (RN), and the radio resources are shared between the UE directly connected to the DeNB and those connected through the RNs [20].

Additionally, 3GPP Release 16 introduces a novel multi-hop wireless architecture called IAB, which enables the NR to utilize a portion of the RAN spectrum for backhaul connections [17]. This advancement makes it feasible to employ NR for wireless backhaul links, connecting central locations to dispersed cell sites as well as linking different cell sites. IAB is applicable to all frequency band where NR is operational. Nonetheless, mmWave spectrum is expected to be the most significant for backhaul links. Additionally, the access link can either operate on the same frequency band as the backhaul link, which is referred to as in-band operation, or use a different frequency band, known as out-of-band operation [15, 18].

The core idea of IAB is to utilize the current 5G access framework for backhaul purposes as well, by efficiently multiplexing access and backhaul across frequency, time, and/or space domains. Although IAB can be supported in both sub-6 GHz and above 6 GHz spectra according to standards, the presence of mmWave spectrum for 5G provides a significant chance to utilize a substantial quantity of new access spectrum that is particularly ideal for IAB. The beam steering capabilities of massive MIMO solutions can be employed to create spatial separation between access and backhaul, thus enhancing spectrum efficiency. This approach enables operators to enhance coverage by deploying denser networks without the immediate need for extensive fiber installation or delaying the costly and complex investment in fiber for backhaul. Consequently, IAB enables and lowers the costs of highly dense deployments, thereby enhancing cellular coverage [15].

7.3.1.1 Multi-hop FWA System with IAB

Initially defined in Release 16, IAB eliminates the need for wired backhaul by functioning as a relay node to extend network access. This approach reduces deployment costs and operational complexity by minimizing the requirement for dedicated backhaul and the intricate maintenance of fiber wiring. In this setup, a gNB with added functionality serves as the IAB-donor, establishing a backhaul link to the 5GC network. An IAB-node can relay connections between other IAB-nodes through an access link to the IAB-donor [5].

To achieve a cost-effective and flexible solution, the IAB concept can be implemented in FWA deployments with multiple hops between APs to enhance spectrum utilization efficiency. In this setup, each residential home connects its CPE to an AP installed on a lamp or outdoor utility pole [14, 20]. Figure 7.3 illustrates a typical multi-hop FWA system using IAB, where only the main BS (MBS) or IAB-donor has a dedicated backhaul connection, usually linked to the core network through fiber or microwave backhaul. In contrast, secondary BSs (SBSs) or IAB-nodes are interconnected through IAB links [14]. At each SBS,

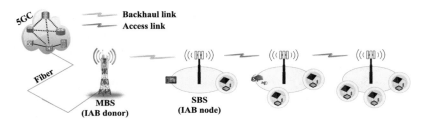

Fig. 7.3 A multi-hop FWA system with IAB

radio resources are allocated dynamically between backhaul and FWA connections. In this setup, SBSs serve consumer CPEs in residential homes and provide backhaul to adjacent SBSs [20].

In IAB networks, backhaul traffic is routed from one SBS to another until it reaches the MBS. This raises the need for efficient backhaul routing, which must take into account factors like SBS load and bandwidth allocation [14]. To facilitate multiple connections and routing options for both backhaul and access, two complementary topologies can be utilized. A spanning tree (ST) topology allows an IAB node to link an IAB-donor through other IAB nodes, supporting multi-hop connections for backhaul. Meanwhile, a directed acyclic graph (DAG) topology provides multiple connections and routes for backhaul redundancy, high availability backup, and load balancing [5].

As MNOs transition from their early 5G market rollouts to expanding network capacity, they encounter the difficulty of implementing high-bandwidth backhaul schemes quickly and cost-effectively. While mm-wave systems offer high-throughput wireless transmission, they are susceptible to blockages caused by mobile objects like vehicles, seasonal changes like foliage, and infrastructure modifications. Therefore, from a dependability standpoint, it is crucial to guarantee that all IAB node can consistently deliver coverage and maintain end-user service, even when active backhaul routes are momentarily disrupted. To facilitate autonomous reconfiguration of the backhaul network without service interruptions or packet loss, 3GPP has standardized topology adaptation. This adaptation can occur due to various factors, including the incorporation of a new IAB node into a current network, the removal of an IAB node, backhaul link congestion detection, degradation of link quality, link malfunction, or other related measures [14].

7.3.1.2 Applications and Implementation Factors

The deployment of mmWave-based 5G exacerbates the challenges of implementing high-bandwidth backhaul schemes by making it significantly harder to secure optimal backhaul solutions. However, the extensive bandwidth available in mmWave and the inherent use of multi-beam technology or massive MIMO systems create a new opportunity for IAB. IAB has the potential to address some of the difficulties service providers encounter when trying

7.3 Advancement of Network Nodes

to offer affordable capacity and coverage solutions. This section explores several IAB use cases that can be leveraged for FWA [15].

Cell Densification

To significantly boost network capacity and deliver exceptional data rates to subscribers, overlaying mmWave 5G small cells on top of a macro cell is an ideal deployment strategy for operators. However, installing fiber for backhaul to support these new small cells can be costly. To address this, operators may opt to use the mmWave spectrum for wireless backhaul due to its substantially broader bandwidth compared to lower-frequency bands. Techniques for mitigating interference, along with potential resource separation methods like time, spatial, or frequency division, can be employed as needed to reduce the effect of sharing the spectrum with backhaul links [15, 21].

As illustrated in Fig. 7.4, IAB can offer a more effective solution for cell densification compared to wired backhaul, depending on the deployment scenario. By wirelessly linking new cells to the backbone network, IAB enhances signal strength within their coverage area, thereby boosting overall network capacity [15, 19].

To demonstrate the advantages of implementing small cells to alleviate the traffic load on a macro BS, consider a scenario with femto BSs–BSs with low transmit power–installed within the coverage area of a high-power macro BS. Specifically, we can analyze the expected number of users served by each femto BS, assuming this number is proportional to the coverage area of the femto BS. Accurately projecting the subscriber count connected to a femto BS is essential, as it directly reflects the volume of traffic offloaded from the macro BS. Thus, the following analysis explores how the anticipated subscriber count within a femto BS's coverage varies with the transmit power of the femto BS.

The coverage area of BS_i, defined as the region where the effective received signal strength indicator (RSSI) of BS_i exceeds that of all other BSs, is specified as [22]

$$\mathcal{A}_i = \bigcap_{j \in \mathcal{N}, j \neq i} \mathcal{A}_i^j, \qquad (7.1)$$

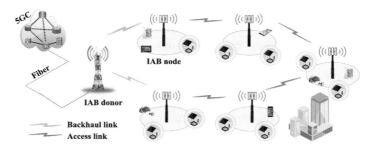

Fig. 7.4 FWA cell densification using IAB

where $\mathcal{N}$ represents the set of BSs, and $\mathcal{A}_i^j$ represents the relative coverage area where the effective RSSI of BS_i exceeds that of BS_j. This area is defined as

$$\mathcal{A}_i^j = \begin{cases} C_{(\chi),d_{i,j}\sqrt{\gamma_{i,j}(\gamma_{i,j}-1)}} & \text{if } \sigma_i P_i < \sigma_j P_j \\ C^c_{(\chi),d_{i,j}\sqrt{\gamma_{i,j}(\gamma_{i,j}-1)}} & \text{if } \sigma_i P_i > \sigma_j P_j \\ \psi \geq \frac{x_i^2 - x_j^2 + y_i^2 - y_j^2}{2} & \text{if } \sigma_i P_i = \sigma_j P_j, \end{cases} \quad (7.2)$$

where $\sigma_i P_i$ is the effective transmit power of BS_i, P_i is the transmit power, σ_i denotes the level of biasing for BS_i, $C_{(x,y),r}$ denotes the region within a circle centered at (x, y) with radius, r and complement, $C^c_{(x,y),r}$; $\chi = x_j + d_{i,j}\gamma_{i,j}\cos(\theta_{i,j}), y_j + d_{i,j}\gamma_{i,j}\sin(\theta_{i,j})$; $\psi = (x, y) | (y_i - y_j) y + (x_i - x_j) x$ and $\theta_{i,j}$, $d_{i,j}$, and $\gamma_{i,j}$ are defined as

$$\theta_{i,j} = \arctan\left(\frac{y_j - y_i}{x_j - x_i}\right)$$

$$d_{i,j} = \sqrt{(x_j - x_i)^2 + (y_j - y_i)^2}$$

$$\gamma_{i,j} = \frac{(\sigma_j P_j)^{-2/\alpha}}{(\sigma_j P_j)^{-2/\alpha} - (\sigma_i P_i)^{-2/\alpha}}. \quad (7.3)$$

Additionally, using a basic PL model featuring a PL exponent α, the projected number of users connected to the femto BS, denoted as $C(P_f)$, is defined as [22]

$$C(P_f) = \frac{\lambda \pi d^2 \left(\frac{\sigma_{mf} P_f}{P_m}\right)^{2/\alpha}}{\left(\left(\frac{\sigma_{mf} P_f}{P_m}\right)^{2/\alpha} - 1\right)^2}, \quad (7.4)$$

where λ represents the user density; P_m and P_f is the transmit powers of the macro and femto BSs, respectively; and σ_{mf} denotes the ratio of the bias factors for the macro and femto BSs.

Assuming $\lambda = 0.001$, $\sigma_{mf} = 1$, and the macro BS transmit power is 43 dBm, the expected number of users, $C(P_f)$, for varying transmit powers of the femto BS can be estimated. Figure 7.5 illustrates that for femto BSs with low transmit powers deployed to alleviate cellular traffic, the expected number of offloaded users, $C(P_f)$, increases as the PL exponent, α, and the distance, d, between the macro and femto BSs increase. This is due to the RSSI of the macro BS diminishes under these conditions.

Addressing Coverage Holes

In 5G networks that operate in high-frequency bands, signal propagation experiences significant diffraction loss and pronounced shadowing, potentially creating areas where signals from cell sites fail to reach–commonly referred to as coverage holes. IAB offers a wireless backhaul solution for new cells added to address these coverage gaps, generally at a lower cost compared to leasing fiber [15, 23]. This use case for addressing coverage holes is illustrated in Fig. 7.6.

7.3 Advancement of Network Nodes

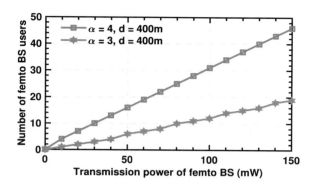

Fig. 7.5 Number of femto BS users versus transmit power of femto BS

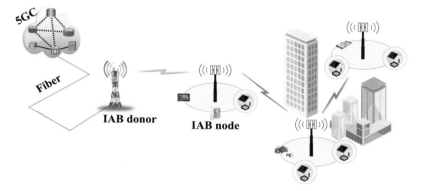

Fig. 7.6 Filling coverage gaps with IAB

IAB can additionally be utilized to broaden coverage into indoor spaces where signals are blocked because of significant penetration loss. By deploying an IAB-node that is positioned to cover both indoor and outdoor areas, coverage can be enhanced without requiring extensive cabling inside the building [15].

On-demand infrastructure

For situations requiring temporary capacity or coverage–such as in stadiums, concert venues, or hazard zones–IAB offers an excellent solution by enabling rapid deployment compared to setting up a dedicated backhaul scheme. IAB nodes can be quickly installed or launched to enhance coverage or service quality in a specific area. Given the temporary nature of these deployments, it is also economical to deactivate the site and IAB when the extra capacity or coverage is no longer needed. IAB nodes can adaptively join or exit the network based on traffic levels and user density, facilitating the flexible creation of an on-demand infrastructure network [15].

Augmenting limited-capacity indoor coverage

In certain indoor settings, such as small businesses and retail stores, the current enterprise Internet backhaul link is often used for radio backhaul. However, the capacity of this backhaul may be constrained, and it may not always meet the precise timing synchronization and latency demands necessary for NR TDD systems. IAB offers an alternative high-speed backhaul alternative to complement or replace the enterprise Internet link. By installing an IAB node on the premises and connecting it to an outdoor IAB-donor for backhaul, high-speed wireless connectivity can be provided inside the inside the facility with minimal capital expenditure [15].

7.3.2 Repeater

While IAB can support the addition of new cells, an RF repeater primarily extends signal coverage, as illustrated in Fig. 7.7. When a repeater is connected to the IAB, it can also function as a gNB to re-transmit the signal. Lightweight repeaters designed to enhance coverage are often cost-effective and straightforward to deploy [5].

7.3.2.1 RF Repeater

An RF repeater functions as a relay node, amplifying and forwarding traffic bidirectionally. It enhances signal strength to extend both the link and coverage but can also increase noise levels, contributing to overall system interference. Consequently, a basic RF repeater may not meet the performance requirements or support essential features such as beamforming and multi-beam operations, which are crucial for NR, especially in the FR2 band.

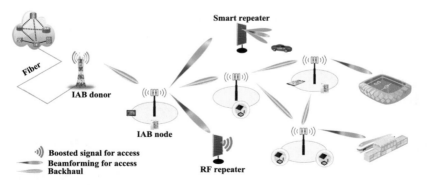

Fig. 7.7 IAB and repeater topology

7.3 Advancement of Network Nodes

7.3.2.2 Smart Repeater

A smart repeater is designed to combine the benefits of RF repeaters and IAB. Unlike traditional RF repeaters, a smart repeater features a significantly advanced antenna array that not only processes side control information but also supports phased array antennas. These antennas are crucial for meeting diverse access needs of UE. With network-controlled capabilities, the smart repeater can utilize multi-beam operation to handle bandwidth-intensive applications using MIMO and to track multiple users through beam scanning and steering.

7.3.3 Intelligent Reflective Surface

The demand for higher data rates in wireless communication is rapidly increasing. This necessitates the development of innovative and cost-effective communication technologies to meet the growing network capacity needs. Massive MIMO technology has the potential to meet the network capacity demands of B5G wireless networks [24–26]. The core concept of massive MIMO involves equipping BSs with dozens, or even hundreds, of antenna elements. This allows them to simultaneously serve multiple mobile terminals using the same time and frequency resources. Although massive MIMO offers numerous benefits, its widespread practical implementation is restricted by the high energy consumption and hardware costs it entails [27, 28]. Furthermore, while mmWave communication gains advantages from massive MIMO through their symbiotic technological convergence, its practical application remains constrained by the limited penetrative ability of mmWave signals, particularly when blockages exist between the mobile device and the BS [29–31].

The concept of IRS has recently emerged and garnered considerable interest as a promising solution for addressing the challenges related to massive MIMO. This technology is also referred to as RIS [32–34]. Unlike a repeater, which amplifies signals, this device simply forwards them. It functions as an intelligent reflecting device that aids in channel management by directing the signal in the desired direction through precise phase shifts applied to the received signal. Additionally, unlike smart repeaters and IAB systems, IRS typically do not include a mobile terminal unit. Instead, they are equipped with a basic controller unit, which can be connected via either a wired or wireless backhaul. Furthermore, IRS systems use a single panel for reflection rather than having separate antenna panels for transmission and reception. Various IRS architectures have been explored in the literature, featuring both active and passive antenna elements. However, from the perspective of 3GPP standardization, since IABs and smart repeaters already incorporate active antenna elements and IRS is envisioned as a cost-effective and energy-efficient device, it is more probable that IRS will consist solely of passive elements [33].

The IRS consists of a flat meta-surface made up of numerous passive reflective elements designed to direct radio signals in a specified direction and in a controlled manner. This method is suitable for applications such as NLoS scenarios indoors, in shopping malls, or in densely populated urban environments [13, 30, 33, 35, 36]. Furthermore, the IRS does not need a power amplifier for transmission, making it an energy-efficient technology. It can passively adjust signal propagation by reconfiguring the phases of its reflective elements via a controller connected to its surface. As a result, the IRS can be used for passive beamforming. Passive beamforming involves adjusting the phases of the IRS without actively powering its antenna elements. This contrasts with active beamforming at the BS, which aims to boost received power and minimize interference for unintended users, thereby improving the network's overall throughput [37]. In practical terms, implementing IRS systems necessitates a significant number of cost-effective phase shifters (PSs) on a surface that can be seamlessly integrated into conventional wireless networks. Consequently, IRS-assisted communication has attracted considerable research attention within the wireless research community in recent years [30, 37, 38].

7.4 Transmission Schemes Convergence

The integration of various transmission technologies, including fiber, FSO communication, and 5G mmWave/NR sub-THz, offers effective solutions for addressing different bottlenecks in communication networks. FSO communication, which transmits optical signals via laser light in free space, has garnered significant interest as a means to address the challenges of short-range 5G wireless communication. As both FSO and 5G technologies rapidly advance, the convergence of FSO with 5G mmWave and 5G NR sub-THz technologies has improved, enabling a decrease in the number of 5G BSs required. This convergence enables high-speed transmission of several tens of gigabits over long distances with FSO, complemented by short-range 5G wireless extensions [3].

For long-haul transport systems, the growing demand for greater transmission capacity across extended distances necessitates advanced solutions. Single-mode fiber (SMF) serves as an excellent transmission medium due to its high bandwidth, low attenuation, and minimal dispersion. Thus, SMF is highly effective in improving the traffic capacity and operational range of FSO-5G mmWave/5G NR sub-THz convergence. This convergence leverages the strengths of optical fiber and FSO/5G wireless technologies, such as the unlicensed bandwidth of free-space links, the vast bandwidth of optical fiber, and the mobility offered by 5G mmWave/5G NR sub-THz wireless transmission [3]. As depicted in Fig. 7.8, a fiber-FSO-5G mmWave/5G NR sub-THz converged system delivers high data rates, high traffic capacity, bidirectional communication, and long-haul transmission. These features provided by the scheme are essential for the successful implementation of FWA.

7.5 Multi-X Connectivity

Managing multiple wireless links is a potential approach to enhance capacity in 5G networks, enabling UE to leverage radio resources from multiple serving cells concurrently and possibly combine bandwidth among them [39]. This can be achieved through Multi-X

7.5 Multi-X Connectivity

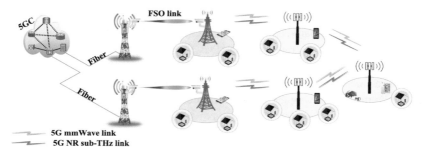

Fig. 7.8 A converged and bi-directional fiber-FSO-5G mmWave/5G NR sub-THz transport systems

connectivity features such as CA and DC [2]. Utilizing Multi-Connectivity (MC) functionality can help mitigate demanding propagation effects like LoS blockage and micromobility. This section introduces the key MC features and discusses their suitability and relevance.

7.5.1 5G Carrier Aggregation and Dual-Connectivity

CA is engineered to enhance the data rate per subscriber by enabling the mobile terminal to connect simultaneously to multiple cells of the serving BS, allowing the UE to function across multiple frequencies concurrently. The emerging paradigm of DC offers an appealing access method in dense, heterogeneous 5G networks, where collaborative schemes and bandwidth sharing are advancing to address growing capacity demands [39]. With DC, the mobile terminal can connect to multiple BSs simultaneously, significantly boosting data rates for cell edge users by allowing simultaneous connections to both LTE and NR [2]. Additionally, DC and CA can be combined, enabling the UE to connect to two BSs and utilize multiple cells in each, thereby drastically increasing the maximum bandwidth available to the mobile terminal.

7.5.1.1 5G Carrier Aggregation

CA is a software feature in RANs and user devices that enables MNOs to merge the capabilities of radio cells operating at different frequency allocations, thereby enhancing the user experience. This technology is crucial for implementing high-performance 4G and 5G networks, offering the ability to aggregate multiple frequency bands for broader cell coverage and higher peak data rates. It significantly boosts coverage, network capacity, and spectral efficiency, making it essential for Beyond 5G (B5G) networks [40].

CA was initially presented in LTE by the 3GPP Release 10 to merge various spectrum segments for higher peak data rates. It proved to be highly effective, becoming an industry standard [40].

When a network combines multiple spectrum segments, one will typically takes on a more prominent role compared to the others. Each segment is known as a component carrier (CC), with the primary component carrier (PCC) being the most important. The others are known as secondary component carriers (SCC). The PCC handles all the UL data, along with both user and control data. The cell serving the PCC is known as the primary cell (PCell), while the cell serving the SCC is referred to as the secondary cell (SCell) [40].

In 5G, the use of various frequency bands with varying slot lengths and numerologies complicates CA. The design of CA must now be sophisticated enough to seamlessly integrate PCells and SCells that may have different slot lengths [40].

7.5.1.2 Dual-Connectivity

5G cellular networks are designed with the aim of handling an increasing number of users and meeting the escalating demand for bandwidth. To accomplish this, two main strategies have been considered. The first strategy involves deploying low-power 5G BSs (such as small-cells, femto-cells, and pico-cells) that overlay the macro-cellular network and utilize its licensed spectrum. This layered architecture facilitates traffic offloading. Specifically, traffic can be offloaded from macro-cells to femto-cells and small cells in this setup. The second strategy involves offloading traffic from the cellular network to alternative networks, such as Low Earth Orbit (LEO) satellites, Wi-Fi, and Low Power Wide Area (LPWA) networks. These 5G HetNets, which integrate various RATs (such as cellular, WiFi, and satellite) or various types of equipment (including macro-cells, small-cells, femto-cells, and pico-cells), have the potential to substantially boost network capacity [41].

Deploying a practical high-rate network with sufficient spectral efficiency necessitates a range of state-of-the-art features. Modern link-level solutions have advanced to near the Shannon limit through the use of advanced Modulation and Coding Schemes (MCS) [39]. With the concept of 5G HetNets and the resulting network densification aimed at enhancing per-user throughput and boosting overall system capacity, the 3GPP has presented DC or MC in HetNets [39, 41]. This approach allows UE to concurrently connect to a macro-cell BS, typically known as the master node, as well as to one or more lower-tier nodes (such as small cells, femto-cells, and pico-cells) and/or heterogeneous nodes (like WiFi AP and satellite), referred to as secondary nodes [39, 41]. DC, in particular, helps meet bandwidth demands and improve data rates while preventing outages. More broadly, MC solutions enhance user session continuity by improving connectivity experience and overall communication reliability [39]. Although DC has been demonstrated to enhance outage performance of mmWave 5G NR systems, it is highly power-intensive due to the energy required to maintain both active and backup links [6].

Like CA, the DC and MC methods allow a UE to concurrently utilize radio resources across multiple component carriers, thereby increasing the bandwidth available to the UE. However, the advantages of MC extend beyond this. Unlike CA, MC/DC allows a UE to connect to different network nodes simultaneously, rather than just one. By using component

7.5 Multi-X Connectivity

carriers from various network nodes, it is feasible to efficiently manage UE mobility without causing disconnections. Additionally, this approach can enhance reliability, particularly for uRLLC, as the same information can be transmitted over multiple channels.

5G NR incorporates several energy-saving techniques, including cross-slot scheduling, bandwidth adaptation, and discontinuous reception (DRX), among others. Among these, DRX is the most extensive mechanism on the UE side, having been tested in 4G LTE and effectively adapted for NR. DRX involves a set of sleep timers that must be optimized to enhance UE energy efficiency. In the context of MC, it is possible to define separate timers for active and backup links, which further boost UE energy efficiency [6].

Additionally, numerous research projects are exploring the capabilities of MC to lower power consumption in 5G networks and enhance spectrum and interference management [41]. Most of these studies aim to enhance both user-cell association and network energy consumption. For example, one study investigates DL and UL Decoupling with DC [39].

7.5.2 Decoupled Uplink and Downlink Access

Generally, spectrum aggregation schemes are directly relevant in the DL, where power accessibility is less of a concern because the evolved Node B (eNB) manages radio transmissions. The UL, however, is often more restrictive as it depends on the user device to perform the transmission tasks. Therefore, extending the allocated bandwidth may not be as advantageous because of the power constraints of the UE. This consideration was also relevant in relation to CA, with various studies examining the likelihood of spectrum aggregation for UL transmissions. In the context of DC, maintaining multiple UL connections can be less energy-efficient for users at the cell edge, mainly because of the higher PL towards the serving cells [39].

As part of a different initiative, 3GPP has introduced the idea of splitting UL and DL to more efficiently offload the macro cells and enhance UL performance. Due to significant transmit power differences between macro and small cells, the cell providing the strongest received power in the DL might not be the one that receives the most power in the UL. Traditional cell association strategies, which depend on DL received power, can lead to suboptimal performance for the UL. Therefore, implementing new cell association policies in heterogeneous networks, designed to enhance energy efficiency and improve user satisfaction across the cell radius, can help achieve better UL rates with greater fairness [39].

The enhanced adaptability afforded by decoupled UL and DL associations offers significant benefits when choosing UL and DL cooperative transmissions or receptions using DC. This flexibility, enabled by the integration of DL and UL Decoupling with DC, represents a major advancement in MC networking. It allows for independent selection of the number and placement of its DL and UL serving cells, taking into account various factors including power limitation, backhaul capacity, and throughput performance. Consequently, spectrum aggregation through DC becomes increasingly effective and adaptable, enabling improved user spectral efficiency [39].

7.6 Adaptive 5G Numerology and Multi-layer Spectrum Management

This section discusses the 5G spectrum designed to support service differentiation, ensuring that users receive the targeted QoE regarding speed, capacity, coverage, and availability.

7.6.1 Multi-layer 5G Spectrum

The 5G spectrum encompasses a broad range of frequency bands to meet various communication objectives. For instance, low-band 5G facilitates long-range communication, whereas high-band 5G delivers superior performance over shorter ranges [42]. To cater to diverse requirements across different 5G use cases, 5G utilizes *high*, *mid*, and *low* frequency bands, utilizing the distinct features of each spectrum segment: frequencies between 2 and 8 GHz (e.g., 3300–4200, 4400–5000, 2500–2690, 2300–2400, and 5925–7125 MHz), along with frequencies below 2 GHz (e.g., 700 MHz) and above 8 GHz (e.g., 24.25–29.5 and 37–43.5 GHz) [43]. These broad bands can handle substantial traffic volumes, which has recently highlighted the potential of FWA [2]. Assigning contiguous broad spectrum bandwidth within each layer simplifies system complexity related to CA, thereby enhancing energy efficiency and reducing network costs [43].

7.6.1.1 5G NR Low-Band
Expanding low-band carriers beyond LTE's 20 MHz bandwidth could enable service provision to the most rural and underserved areas with a minimal number of sites. This approach can also enhance the coverage of Mid-band MU-MIMO deployments at frequencies of 2.5 and 3.5 GHz [2, 44]. Additionally, 5G low-frequency bands (below 2 GHz) will expand the 5G mobile broadband experience to broader areas and deeper indoor spaces. Furthermore, uRLLC and mMTC services will benefit considerably from the superior coverage offered by these low-frequency bands. Accessible low-frequency bands (e.g., 700, 800, 900, 1800, and 2100 MHz) can be utilized for LTE/NR UL spectrum sharing in conjunction with NR in the 3300–3800 MHz range, allowing operators to achieve faster and more cost-effective deployments [43]. Despite its extensive coverage, low-band spectrum is limited in capacity [40].

7.6.1.2 5G NR Mid-Band
Some analysts consider the mid-band spectrum to be the most efficient option for 5G. With carriers ranging from 5 MHz up to 100 MHz in the TDD spectrum, it offers low latency and

7.6 Adaptive 5G Numerology and Multi-layer Spectrum Management

high speeds while providing potential for indoor coverage because of its lower frequency range. This mid-band spectrum also supports MU-MIMO implementations, which enhance the capacity of a single carrier cell and improve per-user Signal-to-Interference-plus-Noise Ratio (SINR), thereby increasing individual user throughput [2].

The 5G mid-band spectrum, spanning from 2 to 8 GHz, is essential for supporting a wide range of 5G use cases across broad areas. Unpaired TDD bands in the ranges of 3300–4200, 4400–5000, 2500–2690, and 2300–2400 MHz offer an optimal balance between extensive coverage and high capacity. Given that eMBB services dominate the early stage of 5G, the mid-band is ideally suited for 5G deployments [43]. It provides greater capacity and bandwidth than low-band spectrum, though its UL coverage is more restricted [40].

7.6.1.3 5G NR High-Band

To achieve increased system capacity and higher user data rates, 5G radio technology will utilize new, often higher frequency bands [45]. Although high-band frequencies do not provide the same level of penetration or coverage distance as mid-band frequencies, they offer extensive spectrum resources (hundreds of MHz) in bands such as 24, 28, 39, and 47 GHz. This enables the delivery of multi-Gbps capacity within a single cell. Additionally, beamforming enhances the SINR for users, which improves the broadband experience [2, 45] and facilitates better coverage at these high frequencies [45].

Massive beamforming at high frequencies produces narrow beams that can be easily redirected when needed. This technique allows signals from several user terminals to be multiplexed concurrently on the same frequency resource but within separate beams, a process commonly known as MU-MIMO. The use of high-gain antennas on both indoor and outdoor terminals enhances the effectiveness of these higher frequencies [45–47]. Thus, high-frequency bands (above 24 GHz) are crucial for delivering additional capacity and achieving the exceptionally high data rates needed for specific 5G eMBB applications in particular high-demand areas (i.e., hotspots) [43]. While these frequencies deliver exceptional peak rates and low latency, they offer less effective UL coverage compared to mid- and low-band frequencies [40].

7.6.2 5G Flexible Numerology

5G NR emphasizes higher frequency bands, operating within the ranges of 410–7125 MHz or 24250–52600 MHz, as illustrated in Table 7.1. The objective of 5G NR is to deliver performance comparable to wired connections, addressing the increasing demand for new devices and bandwidth-intensive applications [40, 42].

To accommodate the diverse 5G deployment setups, ranging from large cells using sub-1 GHz carrier frequencies to mmWave setups with very broad spectrum allocations, 5G offers a flexible numerology defined in 3GPP Release 15. As shown in Table 7.2, this flexible

Table 7.1 Frequency band definitions

Designation	Range	Δf_s (kHz)	B_{max} (MHz)	Applications	References
FR1	Below 2 GHz	15	50/100	For indoor and wide area coverage	[40, 50]
	2–6 GHz	15/30/60	50/100/200	To achieve optimal balance between coverage and capacity	[40, 50]
FR2	Above 6 GHz	60/120/240/480	200/400	For applications requiring high data rates	[40, 50]

B_{max} Maximum bandwidth

Table 7.2 Flexible numerology systems for 5G NR

Frequency band	Duplex mode	μ	S_l (ms)	Δf_s (kHz)	N_{sps}	Cyclic prefix
FR1	FDD	0	1	15	1	Normal
	TDD	1	0.5	30	2	Normal
	TDD	2	0.25	60	4	Normal, Extended
FR2	TDD	2	0.25	60	4	Normal, Extended
	TDD	3	0.125	120	8	Normal
	TDD	4	0.0625	240	16	Normal

N_{sps} Number of slots per subframe S_l Slot length

numerology, denoted as μ, enables a range of subcarrier spacings, Δf_s, which are defined as [48]

$$\Delta f_s = 2^\mu \times 15 \text{ kHz}. \tag{7.5}$$

In contrast to LTE, which utilizes a fixed subcarrier spacing of 15 kHz and a fixed slot length of 1 ms [40], Δf_s in 5G scales by $2^\mu \times 15$ kHz to accommodate various services, considering factors such as QoS, latency requirements, and frequency ranges. Consequently, as numerology μ increases, the number of slots per subframe also increases, resulting in a greater number of transmitted symbols within a given time [49]. The slot length in milliseconds is defined as

$$S_l = \frac{1}{2^\mu}. \tag{7.6}$$

Additionally, subcarrier spacings of 15, 30, and 60 kHz are utilized for lower frequency bands, whereas subcarrier spacings of 60, 120, and 240 kHz are applied to higher frequency bands [49].

7.7 Conclusion

Delivering fixed broadband services involves several challenges, including the substantial costs linked to new infrastructure and the limitations of existing mobile communication bands. To address future traffic demands, it is crucial to explore higher frequency bands that offer larger spectrum allocations. However, these higher frequency bands introduce challenges, particularly related to link budget constraints. Overcoming these challenges requires advanced technical solutions that enhance traffic capacity and extend transmission range. Key strategies, such as Transmission Schemes Convergence, Multi-X Connectivity, and Adaptive 5G Numerology with Multi-Layer Spectrum Management, are vital in making 5G FWA more accessible, cost-effective, and future-ready. These innovations not only expand broadband services but also play a crucial impact in bridging the digital divide, highlighting the importance of ongoing technological progress and strategic planning to overcome deployment challenges.

References

1. Bridging the digital divide: extended-range millimeter-wave 5G Fixed Wireless Access, Ericsson and US cellular, Case study (2022), https://www.ericsson.com/4a9aa7/assets/local/cases/customer-cases/2022/uscellular-bridging-digital-divide.pdf. Accessed 20 Apr 2024
2. Fixed Wireless Access with 5G Networks, 5G Americas, White Paper (2021), https://www.5gamericas.org/wp-content/uploads/2021/11/5G-FWA-WP.pdf. Accessed 21 Apr 2024
3. H.-H. Lu, X.-H. Huang, C.-Y. Li, C.-X. Liu, Y.-Y. Lin, Y.-T. Chen, P.-S. Chang, T. Ko, Bi-directional fiber-FSO-5G MMW/5G new radio sub-THz convergence. J. Lightwave Technol. **39**(22), 7179–7190 (2021)
4. I.A. Alimi, J.J. Popoola, K.F. Akingbade, M.O. Kolawole, Performance analysis of bit-error-rate and channel capacity of MIMO communication systems over multipath fading channels. Int. J. Inform. Commun. Technol. **2**, 57–63 (2013), https://ijict.iaescore.com/index.php/IJICT/article/view/1047/568
5. 5G mmWave Coverage Extension Solutions Whitepaper, GSMA, White Paper (2022), https://www.gsma.com/solutions-and-impact/technologies/networks/wp-content/uploads/2022/12/5gmmWave-coverage-extension-solutions-whitepaper_FINAL-v1.pdf. Accessed 15 Jul 2024
6. V. Beschastnyi, D. Ostrikova, D. Moltchanov, Y. Gaidamaka, Y. Koucheryavy, K. Samouylov, Balancing latency and energy efficiency in mmWave 5G NR systems With multiconnectivity. IEEE Commun. Lett. **26**(8), 1952–1956 (2022)
7. A. Ericsson, L. Falconetti, H. Olofsson, J. Edstam, T. Dahlberg, Closing the digital divide with mmWave extended range for FWA. Ericsson Technol. Rev. **2022**(11), 2–11 (2022)
8. 5G mmWave Coverage Extension Solutions, GSMA, White Paper (2022), https://www.gsma.com/futurenetworks/wp-content/uploads/2022/12/5gmmWave-coverage-extension-solutions-whitepaper_FINAL-v2.pdf. Accessed 21 Apr 2024
9. 5G offers a future-proof platform for FWA growth, Ericsson, Fixed Wireless Access Handbook Extracted version, Insight 6 of 6 (2024), https://www.ericsson.com/4ade15/assets/local/reports-papers/further-insights/doc/fwa_insights_6_offers_extracted.pdf. Accessed 05 Jul 2024

10. K.R. Chaudhuri, E. Cunha Neto, L. Falconetti, R. Fassbinder, S. Guirguis, A. Halder, M. Irizarry, R.D. Patel, N. Saxena, S. Sorlescu, Extended range mmWave for fixed wireless applications, in *2021 97th ARFTG Microwave Measurement Conference (ARFTG)* (2021), pp. 1–4
11. S. Gadhai, R. Budhiraja, Evolution of network nodes: from IAB to IRS, 3GPP, Highlights Issue 06 (2023), https://www.3gpp.org/newsletter-issue-06-may-2023. Accessed 12 May 2024
12. I.A. Alimi, R.K. Patel, A. Zaouga, N.J. Muga, A.N. Pinto, A.L. Teixeira, P.P. Monteiro, *6G CloudNet: Towards a Distributed, Autonomous, and Federated AI-Enabled Cloud and Edge Computing* (Springer International Publishing, Cham, 2021), pp. 251–283. https://doi.org/10.1007/978-3-030-72777-2_13
13. I.A. Alimi, R.K. Patel, A.O. Mufutau, N.J. Muga, A.N. Pinto, P.P. Monteiro, Towards a sustainable green design for next-generation networks. Wirel. Personal Commun. **121**, 1123–1138 (2021). https://doi.org/10.1007/s11277-021-09062-2
14. M.M. Sande, M.C. Hlophe, B.T.S. Maharaj, A backhaul adaptation scheme for IAB networks using deep reinforcement learning with recursive discrete choice model. IEEE Access **11**, 14 181–14 201 (2023)
15. Innovations in 5G Backhaul Technologies: IAB, HFC & Fiber, 5G Americas, White Paper (2020), https://www.5gamericas.org/wp-content/uploads/2020/06/Innovations-in-5G-Backhaul-Technologies-WP-PDF.pdf. Accessed 15 Jul 2024
16. A.H. Jazi, S.M. Razavizadeh, T. Svensson, Integrated access and backhaul (IAB) in cell-free massive MIMO systems. IEEE Access **11**, 71 658–71 667 (2023)
17. Q.-H. Tran, T.-M. Duong, S. Kwon, Load balancing for integrated access and backhaul in mmWave small cells. IEEE Access **11**, 138 664–138 674 (2023)
18. M. Cudak, A. Ghosh, A. Ghosh, J. Andrews, Integrated access and backhaul: a key enabler for 5g millimeter-wave deployments. IEEE Commun. Mag. **59**(4), 88–94 (2021)
19. N. Tafintsev, D. Moltchanov, W. Mao, H. Nikopour, S.-P. Yeh, S. Talwar, M. Valkama, S. Andreev, Analysis of duplexing patterns in multi-hop mmWave integrated access and backhaul systems. IEEE Open J. Commun. Soc. **5**, 5392–5407 (2024)
20. M. Hashemi, M. Coldrey, M. Johansson, S. Petersson
21. I. Ajewale Alimi, N. Jesus Muga, A.M. Abdalla, C. Pinho, J. Rodriguez, P. Pereira Monteiro, A. Luís Teixeira, *Towards a Converged Optical-Wireless Fronthaul/Backhaul Solution for 5G Networks and Beyond* (John Wiley & Sons, Ltd, 2019), pp. 1–29, https://onlinelibrary.wiley.com/doi/abs/10.1002/9781119491590.ch1
22. W. Lee, R. Schober, Analysis of coverage in heterogeneous cellular networks. IEEE Commun. Lett. **20**(6), 1211–1214 (2016)
23. C.K. Anjinappa, S. Güvenç, Coverage Hole detection for mmWave networks: an unsupervised learning approach. IEEE Commun. Lett. **25**(11), 3580–3584 (2021)
24. F. Rusek, D. Persson, B.K. Lau, E.G. Larsson, T.L. Marzetta, O. Edfors, F. Tufvesson, Scaling Up MIMO: opportunities and challenges with very large arrays. IEEE Signal Process. Mag. **30**(1), 40–60 (2013)
25. E.G. Larsson, O. Edfors, F. Tufvesson, T.L. Marzetta, Massive MIMO for next generation wireless systems. IEEE Commun. Mag. **52**(2), 186–195 (2014)
26. T.L. Marzetta, Noncooperative cellular wireless with unlimited numbers of base station antennas. IEEE Trans. Wirel. Commun. **9**(11), 3590–3600 (2010)
27. S. Buzzi, I. Chih-Lin, T.E. Klein, H.V. Poor, C. Yang, A. Zappone, A survey of energy-efficient techniques for 5G networks and challenges ahead. IEEE J. Select. Areas Commun. **34**(4), 697–709 (2016)
28. S. Zhang, Q. Wu, S. Xu, G.Y. Li, Fundamental green tradeoffs: progresses, challenges, and impacts on 5G networks. IEEE Commun. Surv. Tutor. **19**(1), 33–56 (2017)

29. X. Tan, Z. Sun, D. Koutsonikolas, J.M. Jornet, Enabling indoor mobile millimeter-wave networks based on smart reflect-arrays, in *IEEE INFOCOM 2018—IEEE Conference on Computer Communications* (2018), pp. 270–278
30. H. Ur Rehman, F. Bellili, A. Mezghani, E. Hossain, Joint active and passive beamforming design for IRS-assisted multi-user MIMO systems: a VAMP-based approach. IEEE Trans. Commun. **69**(10), 6734–6749 (2021)
31. L. Yashvanth, C.R. Murthy, On the impact of an IRS on the out-of-band performance in sub-6 GHz & mmWave frequencies. IEEE Trans. Commun. 1–1 (2024)
32. C. Liaskos, S. Nie, A. Tsioliaridou, A. Pitsillides, S. Ioannidis, I. Akyildiz, A new wireless communication paradigm through software-controlled metasurfaces. IEEE Commun. Mag. **56**(9), 162–169 (2018)
33. 5G mmWave Deployment Best Practices, GSMA, White Paper (2022), https://www.gsma.com/solutions-and-impact/technologies/networks/wp-content/uploads/2022/10/5G-mmWave-Accelerator-Deployment-Best-Practices-FINAL-UPDATE-JAN-2022.pdf. Accessed 05 Jul 2024
34. S. Hu, F. Rusek, O. Edfors, Beyond massive MIMO: the potential of data transmission with large intelligent surfaces. IEEE Trans. Signal Process. **66**(10), 2746–2758 (2018)
35. H. Choi, A.L. Swindlehurst, J. Choi, WMMSE-based rate maximization for RIS-assisted MU-MIMO systems. IEEE Trans. Commun. **72**(8), 5194–5208 (2024)
36. S. Gong, C. Xing, P. Yue, L. Zhao, T.Q.S. Quek, Hybrid analog and digital beamforming for RIS-assisted mmWave communications. IEEE Trans. Wirel. Commun. **22**(3), 1537–1554 (2023)
37. Q. Wu, R. Zhang, Intelligent reflecting surface enhanced wireless network via joint active and passive beamforming. IEEE Trans. Wirel. Commun. **18**(11), 5394–5409 (2019)
38. S. Pala, O. Taghizadeh, M. Katwe, K. Singh, C.-P. Li, A. Schmeink, Secure RIS-assisted hybrid beamforming design with low-resolution phase shifters. IEEE Trans. Wirel. Commun. **23**(8), 10 198–10 212 (2024)
39. M.A. Lema, E. Pardo, O. Galinina, S. Andreev, M. Dohler, Flexible dual-connectivity spectrum aggregation for decoupled uplink and downlink access in 5G heterogeneous systems. IEEE J. Sel. Areas Commun. **34**(11), 2851–2865 (2016)
40. What, Why and How: the Power of 5G Carrier Aggregation. Ericsson, Tech. Rep. (2021), https://www.ericsson.com/en/blog/2021/6/what-why-how-5g-carrier-aggregation. Accessed 22 Apr 2024
41. T. Sylla, L. Mendiboure, S. Maaloul, H. Aniss, M. A. Chalouf, S. Delbruel, Multi-connectivity for 5G networks and beyond: a survey. Sensors **2**(19) (2022), https://www.mdpi.com/1424-8220/22/19/7591
42. What Is 5G NR? 5G New Radio Standard Explained, Celona, Position Paper, https://www.celona.io/5g-lan/5g-nr. Accessed 23 Apr 2024
43. 5G Spectrum Public Policy Position, Huawei, Position paper (2020). Accessed 23 Apr 2024
44. S. Mukherjee, M.S. Khan, A. Kumar Reddy Chavva, Optimizing near-field XL-MIMO communications: advanced feedback framework for CSI. IEEE Access **12**, 89 205–89 221 (2024)
45. K. Laraqui, S. Tombaz, A. Furuskär, B. Skubic, A. Nazari, E. Trojer, Fixed wireless access on a massive scale with 5G. Ericsson Technol. Rev. **94**, https://www.ericsson.com/assets/local/publications/ericsson-technology-review/docs/2017/2017-01-volume-94-etr-magazine.pdf
46. S. Adda, T. Aureli, S. D'Elia, D. Franci, N. Pasquino, S. Pavoncello, R. Suman, Enhanced methodology to characterize 3-D power monitoring and control features for 5G NR systems embedding multi-user MIMO antennas. IEEE Trans. Instrum. Meas. **72**, 1–9 (2023)
47. K.R. Jha, N. Rana, S.K. Sharma, Design of compact antenna array for MIMO implementation using characteristic mode analysis for 5G NR and Wi-Fi 6 applications. IEEE Open J. Antennas Propag. **4**, 262–277 (2023)

48. A.P.K. Reddy, M.S. Kumari, V. Dhanwani, A.K. Bachkaniwala, N. Kumar, K. Vasudevan, S. Selvaganapathy, S.K. Devar, P. Rathod, V.B. James, 5G new radio key performance indicators evaluation for IMT-2020 radio interface technology. IEEE Access **9**, 112 290–112 311 (2021)
49. S. DeTomasi, 5G flexible numerology—defining what it is and explaining why you should care. Keysight, Tech. Rep. (2018), https://www.keysight.com/blogs/en/inds/2018/09/07/5g-flexible-numerology-defining-what-it-is-and-explaining-why-you-should-care. Accessed 23 Apr 2024
50. A. Dogra, R. K. Jha, and S. Jain, A survey on beyond 5G network with the advent of 6G: architecture and emerging technologies. IEEE Access **9**, 67 512–67 547 (2021)

Conclusion 1

The COVID-19 pandemic highlighted the essential need for reliable high-speed broadband to facilitate remote work and online learning-trends that have persisted beyond the pandemic. To meet this continuous demand, home broadband must be robust, adaptable, reliable, and secure. Despite the expansion of both fixed and mobile broadband, many households remain underserved. FWA offers an effective solution to this issue by utilizing the extensive infrastructure, broad cellular spectrum, and global reach of mobile technologies such as LTE and 5G NR.

5G technology represents the culmination of a decade of research, standardization, and industry collaboration. It offers a chance for true network convergence, enabling the same technology and infrastructure to support next-generation IoT, MBB, and FWA. Technological advancements in 5G NR, such as increased spectrum allocation, beamforming, improved terminals, optimized media distribution, network virtualization, new frequency bands, and support for higher numerology, are set to usher in an era of boundless possibilities and comprehensive connectivity. Additionally, features like MU-MIMO are particularly beneficial for FWA, using the geographical separation of users to create multiple, more efficient beams, thereby increasing spectral efficiency and enhancing peak rates for home CPEs. Furthermore, new capabilities like network slicing and service differentiation will enhance 5G-based FWA offerings from an e2e perspective. Advances in mobile backhaul and fronthaul, network virtualization, and network programmability are also significantly strengthening the FWA concept.

The radio signal processing stack in 5G NR operates as a sequential chain of functions. These functions can be decomposed and separated with defined interfaces to achieve disaggregation. Both HLS and LLS alternatives offer their own advantages and disadvantages. The ideal split is determined by various technical and business factors, including network topology, fiber availability, user count, and service volume. This split in the 5G RAN architecture is crucial for developing FWA solutions, as it allows for the dynamic placement of functions and the flexibility needed to adapt to future demands for capacity, latency, and evolving applications like AR and VR in residential settings.

© The Author(s), under exclusive license to Springer Nature Switzerland AG 2025
I. Alimi, *5G Fixed Wireless Access*, Synthesis Lectures on Communications,
https://doi.org/10.1007/978-3-031-77539-0_1

Planning coverage for traditional wireless networks is a complex and resource-intensive task that often requires substantial time, manpower, and expensive equipment to ensure comprehensive coverage and prevent gaps. Poor planning can result in subpar coverage quality or complete black spots, leading to high costs and prolonged resolution times. However, incorporating AI into coverage planning tools has revolutionized the process. AI-based methods enhance test points, making the planning process more accurate and efficient, and greatly improving the effectiveness of coverage tools. This approach not only streamlines planning but also allows for more precise CPE installation, resulting in better network performance and increased customer satisfaction.

5G MNOs are ideally situated to address the varied broadband requirements of residential customers by maximizing their current resources. For instance, 5G operating on low-frequency bands can deliver sufficient broadband service to most rural and underserved regions with a limited number of sites. The mid-band spectrum strikes an effective balance between capacity and coverage range. Spectrum carriers spanning from 5 to 100 MHz can provide high-speed broadband with indoor coverage. Furthermore, MNOs can utilize both underused and undeployed spectrum bands to meet the increasing residential broadband demand.

Additionally, well-designed CPE is vital for delivering excellent customer service and optimizing existing network configurations. The placement of CPE is a critical element in an FWA operator's strategy, as it greatly influences customer adoption and the profitability of the FWA business. Indoor CPEs are often favored for their ease of self-installation and suitability for FWA products operating on mid or low bands. Conversely, outdoor CPEs, with their high-gain antennas, address the propagation challenges of higher mmWave bands, extending the cell range in rural areas, although they necessitate professional installation on the home's exterior. Window-mounted solutions are also becoming popular as a self-install option, providing the benefits of an external antenna with an improved link budget.

Both rural and urban areas present substantial opportunities for 5G-based residential FWA services. Historically, rural regions have lagged behind urban areas in broadband adoption, but 5G-based FWA is poised to change that by offering a promising solution to bridge the digital divide. Although the specifics of any FWA deployment depend on the context, research indicates that 5G-based FWA is a viable option to meet the advanced future service needs of homes and SMEs in various environments worldwide.

There is no one-size-fits-all solution for the transport network, so operators should carefully consider the available options and their respective advantages and disadvantages. Notably, 5G fixed wireless technology can coexist with other high-speed broadband delivery methods like hybrid fiber cable, xDSL, and FTTx, providing service providers with a range of options depending on geography, customer needs, and commercial viability. This enables 3GPP-based MNOs to quickly and efficiently offer viable broadband solutions to rural and underserved areas. 5G-based FWA can enhance Internet speeds and provide a consistent experience across both rural and urban communities, offering residential consumers-especially in rural areas-greater choice of broadband providers and helping to close the digital divide.